BROUGHTON

FROM WELLINGTON TO AIRBUS

The Broughton Wellington: The Vickers Wellington Mk.IC R1333 Second World War bomber that was subscribed by the 'Workers and Co-operators' of the Vickers-Chester Shadow Factory built in 1939 at the Flintshire, North Wales village of Broughton, near Chester, and ceremoniously presented to the Royal Air Force in November 1940. Wellington production there totalled 5,540 aircraft in the six years between August 1939 and September 1945 and peaked at 130 aircraft per month.

Global-Girdling Airbus Wings Upon Wings: Loading the first wing structure for the newest derivative of the globally successful pan-European Airbus airliner family – the 380-seat A340-600 Series – at the BAE SYSTEMS Airbus UK manufacturing centre at Broughton in mid-April 2000. The carrier is one of the Airbus 'in-house airline' fleet of five custom-built A300-600ST Super Transporter 'Beluga' aircraft – which also incorporate Broughton-made wings. This wing structure was then flown to the DaimlerChrysler (now Airbus Deutschland) plant at Bremen, Germany for equipping and after to the Airbus final assembly centre at the Aerospatiale-Matra (now Airbus France) factory at Toulouse, south-west France. More than 2,600 sets of Airbus wings have been completed at Broughton since 1971.

BROUGHTON

FROM WELLINGTON TO AIRBUS

Norman Barfield

Dedicated to the memory of
Jeffrey 'Monty' Montgomery

A long-standing Vickers professional engineer and caring friend to many, he spent his early years at Broughton as his father was the first Works Manager there.

Also to those many thousands whose exceptional efforts over the past six decades have built the Broughton Factory into the world-class facility that it is today.

TEMPUS

First published 2001
Copyright © Norman Barfield, 2001

Tempus Publishing Limited
The Mill, Brimscombe Port,
Stroud, Gloucestershire, GL5 2QG

ISBN 0 7524 2130 1

Typesetting and origination by
Tempus Publishing Limited
Printed in Great Britain by
Midway Colour Print, Wiltshire

Cover Illustration:
The Broughton-Made Airbus Wing Family Millennium Parade
Airbus UK Broughton-made wings fly on every member of the successful Airbus jet airliner family
which is now winning around half of all mainstream airliner orders. This rising echelon 'Airbus Family
Flight' of all eight Airbus aircraft types currently in production symbolizes the highest achievement of
the six-decade cavalcade of Broughton-made products. It was photographed at the Paris air show, Le
Bourget, in June 1999 to celebrate the 30th anniversary of the launch of the initiating Airbus A300,
the world's first twin-aisle, twin-engined jetliner, in 1969. This eminent aircraft manufacturing centre
has always been the exclusive supplier of Airbus wings and this distinctive line-up has become the
inspiration for Broughton and all the constituents of the Airbus airliner programme for the new
millennium. The formation comprises (top to bottom): the A319 Airbus Corporate Jetliner, A320 and
A321 that comprise the single-aisle A320 Airbus family; the twin-jet A310-300, A300-600 and
A330-200 twin-jets and the four-engined A340-300 that comprise the twin-aisle wide-bodied family;
together with one of the fleet of five specially-adapted A300-600ST Super Transporter ('Beluga') inter-
site airframe component airlifters. Fortuitously, this signal event also coincided with the 60th
anniversary of the first flight of the first Broughton-made Vickers Wellington bomber, in August 1939,
for the large-scale production of which the plant was originally built. It thereby effectively encapsulates
Britain's, and one of the world's, most outstanding aircraft production records.

Contents

The Shadow Factory Scheme

The rise to power of Adolf Hitler, with his accession to the German Chancellery in 1933 and his repudiation of the terms of the Versailles Treaty of 1919, resulted in the British Government abandoning its belief that disarmament was the road to peace. They had to face the reality that Germany was building up its arsenal in readiness for another war.

The situation quickly came to a head at the League of Nations Disarmament Conference of October 1933 in Geneva when the German delegation precipitously walked out. Despite strenuous efforts to revive it, their action ultimately resulted in the Conference being abandoned in June 1934.

The evident alarm this caused, together with widespread rumours of greatly increased military aircraft production in Germany, led the British Prime Minister, Stanley Baldwin, to announce a substantial Expansion Programme for the Royal Air Force on 19 July 1934. The primary objective was to increase aircraft production in order to achieve parity with the German *Luftwaffe* by no later than April 1939.

At the same time, the aircraft industry was introducing radical changes in design – more powerful engines and more potent armaments – and grappling with the increased time taken to ready these new types for operational service. The manufacturing capacity of the industry was grossly insufficient to meet the perceived needs of the RAF and therefore required a correspondingly radical solution.

So it was that the Government conceived the 'Shadow Factory Scheme'. This consisted principally of utilizing related manufacturing capacity outside the aircraft and aero-engine industries. Primarily, it included the expanding motor vehicle industry, which was by then the national leader in large-scale mechanical engineering, metallic fabrication and mass production techniques, with the major motor manufacturers 'shadowing' the professional aviation industry.

The function of these additional factories would be to produce in quantity any type of airframe or engine that had been officially frozen in design. This, in turn, would allow the parent firms to develop new or improved types and, in reasonable time, to establish large-scale production at the shadow factories, enabling production to keep abreast of operational requirements.

Accordingly, the Air Ministry invited six major motor manufacturers to participate. The Government funded the new factories but the firms themselves supervised the construction and management, including the recruitment and training of the necessary labour force.

The locational criteria were that the factories should be near to populous areas from which the requisite labour force could be drawn, have ease of road and rail access, and be on or near existing aerodromes to facilitate flight testing and delivery (or, if not, to include the construction of a new one). They should be geographically dispersed to prevent stoppage in the event of a successful enemy air attack on one of them and hence be situated out of range of enemy bombers where possible.

Several such large units were thus established in the Midlands and the Northwest of England, but away from the well-established aircraft factories with locations, mainly in the South, which were well-known to the enemy, and most of them were managed by the motor industry. However, those for the Vickers Wellington bomber were not. The project had been seriously considered at the big Nuffield plant at Castle Bromwich, Birmingham but, in the event, Spitfire fighters were built there instead, also under the aegis of Vickers.

Vickers-Armstrongs had planned a large dispersal factory of its own for the Wellington at Broughton, near Chester – the site of which, and its subsequent story to date, is the subject of this book (together with a second autonomous plant near Blackpool which is also included here as a directly related cameo). Both plants ultimately became Government-owned, Vickers-managed, units within the Shadow Factory Scheme.

Introduction

Six decades on from its original construction as a 'Shadow Factory' for large-scale production of the Wellington bomber (one the few aircraft types that continued in production throughout the Second World War), and three decades on from the formation of the pan-European Airbus Industrie (AI), the present Airbus UK factory at Broughton, Flintshire, North Wales is the exclusive manufacturing centre for the wing structures of the globally-successful Airbus airliner family. It is widely acclaimed as one of the most productive and efficient aerostructure plants anywhere in the world.

Passing successively from Vickers-Armstrongs Airbus to de Havilland, Hawker Siddeley, British Aerospace, BAE SYSTEMS and now Airbus UK, this huge facility has delivered nearly 9,500 military and civil aircraft of fifteen distinctly different types. They range from the famed Wellington and Lancaster of wartime vintage, through to the World-leading Comet jetliner and 125 business jet. The factory has also produced over 2,600 sets of Airbus airliner wings during the second half of this time period: a production record unparalleled by any other British aircraft factory.

Although never having had an autonomous corporate identity of its own, the Broughton factory's exceptional production record and provenance have meant it has gained influence and a reputation equivalent to that of the founding British aircraft firms that featured so strongly in its early years.

Essentially beginning as a complete aircraft assembly plant for the Wellington, although much component and detail support work has also been introduced over the years, the Broughton plant has always relied heavily on external supply. In the Vickers years, this consisted of an exclusive network of 500 suppliers in the Northwest and the Midlands (and from the several Avro wartime plants in the Manchester area for the Lancaster). During the de Havilland era, major components were imported from the DH sites at Hatfield, Portsmouth, Christchurch and Downsview, Canada; also from Fairey Aviation at Stockport, Saunders Roe at Eastleigh, Southampton and Cowes, Isle of Wight. In the Hawker Siddeley and British Aerospace years, this circle was widened even further to include Woodford and Chadderton, near Manchester, Brough, Yorkshire, and Hamble, near Southampton.

The massive build-up of Airbus wing construction in the British Aerospace/BAE SYSTEMS regimes has embraced many other UK and overseas sources, notably including Weybridge, Filton, Prestwick, Australia, the USA and 400 other suppliers throughout the country. Broughton itself has simultaneously become a major component supplier to the continental-based Airbus final assembly centres at Toulouse and Finkenwerder, Hamburg.

The Broughton factory is now fully modernised and advancing into the new millennium with much energy to fulfil the global Airbus order book. It has good prospects with regards to supporting the continually expanding Airbus family lineage and is not only the pride of British aerostructures manufacture but an acknowledged model of quality and efficiency for the rest of the aerospace world.

Genesis

Emerging as a key element of the British Government's 'Shadow Factory Scheme', the decision to build a new aircraft production factory was originally taken in September 1936. When preparations were being made to put the new Vickers geodetic ('basketweave') bomber, the Wellington, into full-scale production, after the prototype had first flown at the company's headquarters plant at Weybridge, Surrey, on 15 June of that year, it was realised that the factory facilities there would not be able to cope alone with the size of the anticipated orders for the re-designed production version.

Erection of the steelwork of the main factory area which, when completed in September 1939, was traversed throughout the unsupported roof girders by a unique and powerful overhead rapid-transit crane system.

Sir Robert McLean, Chairman of Vickers (Aviation) Ltd, approached the British Government with plans for an additional factory. Gordon ('Monty') Montgomery was a Vickers-Weybridge veteran, a contemporary of Tommy Sopwith at Brooklands and member of the Royal Flying Corps. He had notably lead the team that assembled the historic Alcock and Brown Vickers Vimy of transatlantic fame in Newfoundland in 1919 and supervised the building of the Vickers B9/32 Wellington prototype in 1936 and the Wellesleys of the RAF's record-breaking Egypt-Australia Long Range Flight in 1937/1938.

Towards the end of 1937, Montgomery ostensibly took his family on holiday in North Wales and Scotland. The real reason for these excursions, however, was to assess the options of building a Wellington shadow factory near Chester, in the north-east of England or in Inverness. In the event, the former was chosen because the latter was considered too remote and subject to unacceptable weather conditions for continuous winter operations. The Chester site was one of the few areas in North Wales that could accommodate an airfield with a reasonably clear year-round test flying area.

Building A Bomber Birthplace

Whereas most of the shadow factories were managed by the motor industry, the Vickers Wellington was not, in the event, built under the motor industry scheme. Originally intended as Vickers-Armstrongs' own installation, and solely devoted to production of the Wellington, the decision was taken in 1938 for the Chester factory to be Government-owned but leased to, and managed by, that company and sharing the adjacent aerodrome with the Royal Air Force.

The intention was that the factory would be a 'parent works' – i.e. planned as a huge mass-production assembly plant only, laid out and built on the concept that the manufacture of all components would be subcontracted – a method which was considered to be eminently suitable for such an aircraft as the Wellington. This was quite different from Weybridge, where aircraft were built complete, from raw material to the finished product. The same concept applied to the second Vickers-managed Wellington Shadow Factory at Blackpool on the Fylde coast of Lancashire which was, in many respects, a modernized duplicate of Weybridge.

The original Vickers-Armstrongs wartime Shadow Factory and Royal Air Force Aerodrome, as they were upon completion in September 1939 (but with the five small hangars in the foreground early in the de Havilland tenure), facing the Hawarden aerodrome main runway in a Northeasterly direction. This completely new Wellington bomber assembly production facility was completed and operational in only nine months from the initial ground-breaking in mid-December 1938.

As designed, the Chester factory had no 'back shops', such as fitting, machining etc., other than millwrights. Initially, there were no finished parts stores and it was intended that co-ordinated stocks would be delivered on a daily basis directly to the production line (an early example of the modern-day 'just-in-time' and 'line-side supply' practices). The plant did not continue in this fashion. As the tempo of production built up, large stocks of parts did have to be accommodated on site.

The ground-breaking and construction of the new factory began on 14 December 1938, the civil engineering contractor being Sir William Arrol and Son of Clydeside, Scotland. The builders of Tower Bridge and the Forth bridge had also constructed the nearby Queensferry Bridge in 1926, which spanned the River Dee to the Wirral peninsular, traversed daily by many of the factory employees over the past sixty years, and is still registered in the portfolio of Langley Holdings, the diversified engineering group of Redford, Nottinghamshire. The completed factory became fully operational in September 1939. The distinctive feature of the layout was the huge open assembly area, some 1,200ft long, at the end of which was an unsupported roof area of 650ft square. This gave it the largest unsupported single roof span and floor area of any aircraft factory in Europe – 1.5million sq.ft with a main assembly area of around 1million sq.ft.

Adjoining land was purchased by the Air Ministry during construction from the Hawarden estate, where the supporting aerodrome was built – originally for the RAF – by Bernard Sunley & Co., who designed the Birmingham Elmdon airport amongst others (still operating today as Sunley Holdings).

What's In A Name?

At this point, it is necessary to clarify the sometimes-confusing matter of the name of the factory. Predominantly known throughout most of its six decades as 'Chester', because of the historic Roman walled city four miles to the east in the English county of Cheshire being the nearest well-known landmark. The actual geographical location is on the outskirts of the village of Broughton (pronounced 'Brawton'), about three-and-a-half miles to the Southwest of Chester; it is, in fact, a mile inside the Welsh border in the county of Flintshire. Hence, it is by this latter name that the factory has often also been known, especially locally. To add to this confusion, it has also been known as 'Hawarden' (pronounced 'Harden'), the name of better known small town from which the original supporting RAF aerodrome took its name.

The unique and distinctive geodetic ('basketweave') structure of the Wellington airframe: 8,946 of these aircraft were built at the Vickers-Chester and -Blackpool shadow factories (5,540 and 3,406 respectively) during the Second World War.

It is now the policy of Airbus UK to use the name 'Broughton' exclusively. This is principally in deference to the recent Welsh political devolution, being the only aircraft manufacturing facility in Wales (around 60 per cent of the workforce traditionally being Welsh). All three names appear in this book as applicable in their relevant contexts, time periods and usage.

The Vickers Bastion Of Big Bomber Production

The initial order for 180 Wellingtons was placed with Vickers-Weybridge in August 1936 (only two months after the prototype first flight). In October 1937, prior to setting up the Chester factory, a contract for a 100 Bristol Pegasus-engined Mk. Is was assigned to the Gloster Aircraft Company. These were to be followed by a second batch of 100 of the projected Rolls-Royce Merlin-engined Mk. II. However, by early 1939, when the Chester factory was nearing completion, and a further contract for 750 aircraft was placed with Vickers in May of that year, the Gloster contracts were rescinded by the government and all raw materials and finished and unfinished parts were transferred to the new plant. Sometime later, an order for 64 Wellingtons that had originally been placed with Sir W. G. Armstrong Whitworth Aircraft of Coventry, was also transferred to the Blackpool factory

Preparations were being made by de Havilland in 1940 for the Mosquito to replace the Airspeed Oxford in production at its parent plant at Hatfield, Hertfordshire. As the Mosquito was not yet a certainty, de Havilland was also approached later that year to take on the building of the Vickers Wellington, rising to a rate of 300 units per month, and this they agreed to do. As with the Vickers shadow factories at Chester and Blackpool, it involved finding a suitable site with airfield, supporting labour, and subcontractors. Although the Wellington option did not materialise, this is how the DH 'Second Aircraft Group', centred at Leavesden, near Watford, originated.

Gordon Montgomery was given the daunting job of assembly manager to get the Chester Wellington production started from scratch. Having been closely involved in making the Meccano-like geodetic construction a practical proposition at Weybridge, only a handful of key Weybridge people had to be transferred with him to Chester. Together, they began to construct

Wellington production at the Vickers-Chester Shadow Factory.

the first aircraft on 3 April 1939, in a temporary Bellman-type hangar on the edge of the site near the Glynne Arms public house, while the main factory was being built. Using parts supplied by Weybridge and various subcontractors, this first Chester-built machine, a Wellington I L7770, was completed four months later and first flown on 2 August 1939. Leaving the Hawarden airfield initially for Weybridge for production testing, it was delivered to No.99 Squadron RAF in early September, just as the Second World War was starting. The new factory became fully operational in September 1939, only nine months from the start of construction, and by the end of December 1939 they had completed five aircraft. Although local subcontractors supplied most parts for the factory, as intended, there was still considerable reliance on support from Weybridge. By the first anniversary of the start of production, twenty-one aircraft had been built – a remarkable achievement as 98 per cent of the workforce were still only semi-skilled.

Paralleling the rapid progress in aircraft delivery was the corresponding build-up of the workforce. On 1 January 1940, there were only 697 workpeople but by the end of September 1943 this number had increased to 5,546 with a further 1,500 on the staff. Significantly, 68 per cent of the assembly workers were women and at the peak of production the head foreman of the final assembly area was a young woman. This was largely because the construction of the fabric covering of the Wellington airframe was akin to tailoring and therefore represented a major contribution for which female labour was regarded to be ideally suited.

The devastating air raid on the geographically vulnerable Vickers-Weybridge factory on 4 September 1940, and the resulting severe (if commendably brief) disruption of the well-established Wellington production line there, was an early vindication of the Shadow Factory dimension.

Unfortunately, Gordon Montgomery died in 1941, only forty-six years old, but not before the Chester production rate had been built up to more than thirty aircraft a week.

To help maintain the overall Wellington production output at the three factories, it became necessary for each to open a satellite assembly plant – for Weybridge at Windsor Great Park (Smith's Lawn), for Chester at Byley/Cranage and for Blackpool at Stanley Park aerodrome. Opened in 1941, the Chester satellite assembly line was actually located at Byley, near Middlewich in mid-Cheshire, about twenty-five miles east of the Chester plant. It was some fields away from the Cranage aerodrome, which was just to the north-east of the flight shed, which was only used for the flying. However, this combined facility was always known as 'Cranage'. The first machine was completed there in September of that year and several hundred aircraft were subsequently built there.

From the outset, the Chester/Cranage Wellington production grew rapidly, peaking at a monthly output of 130 aircraft (although there is no known breakdown between the two assembly lines), concentrating on bomber and crew training marks, and reaching a combined

The Broughton / Hawarden Shadow Factory / RAF aerodrome complex in the later stages of the Second World War with large numbers of Vickers Wellington and Avro Lancaster bombers awaiting delivery.

total of 5,540 aircraft when production ceased in September 1945. This was only 380 less than the combined total built at Weybridge and Blackpool (5,920). The Chester and Blackpool shadow factories had between them built nearly four times the number built by the long-established parent factory at Weybridge (2,514) but both shadow factories had, of course, been purpose-built and laid out more spaciously for high rates of production from the start.

Together, the three factories sustained Wellington production throughout the War, making it Britain's most prolific two-motor bomber aircraft programme with a total of 11,460 aircraft produced between them. They made the Wellington one of the most successful aircraft programmes ever – and serving with every command of the RAF, except fighter.

As the strategy of the Royal Air Force changed to heavy bomber air raids on key German industrial targets, in March 1943 Government instructions were received to taper off the Wellington production (orders for more than 600 aircraft were cancelled) and for the Chester factory to join the six-company Avro Lancaster Production Group. An order was then placed with Vickers-Armstrongs on 5 April 1943 for 500 Lancaster B.Mk.I four-engined heavy bombers to be built there.

By September 1944, the Lancaster order book at Chester stood at 680 aircraft and a further 840 of an improved type, the Lincoln. However, the end of the War in Europe in May 1945, and in the Far East in August, meant a drastic curtailment. Only 235 Lancasters were actually built between June 1944 and June 1945. The aircraft production programme at Chester was ultimately completed after twenty-three Lancasters and eleven Lincolns had been assembled from components transferred from the Metropolitan-Vickers Electrical Co. factory at Trafford Park, Manchester, between 15 June and 13 August 1945 (and generally included in the total Chester output). These last aircraft were subsequently flown to Sir W. G. Armstrong Whitworth Aircraft at Coventry late in 1946 for operational conversion to Lancaster B.Mk.I F.E. (Far East) standard (as had been around fifty-five of the earlier Chester-built aircraft).

Near to the completion of the final Ministry of Aircraft Production (MAP) orders, the Chester plant began to slow down, facing closure during the latter months of 1945. Fortuitously, however, an unexpected and substantial new work programme was to materialise just in time to avert this potentially serious eventuality – the construction of 11,250 prefabricated domestic dwellings.

Large-scale construction of pre-fabricated aluminium houses for which both the Chester and Blackpool factories were adapted at the end of the Second World War in 1945. Together they produced 22,500 of these ingenious constructions during the ensuing three years.

The Blackpool Sibling

Although not directly part of the subject of this book, it is relevant to include the parallel contribution made by the second Vickers Wellington Shadow Factory, located at the Squires Gate airport near Blackpool. As in the case of Chester, this plant was owned by the government and operated by Vickers-Armstrongs and had a second assembly line established at nearby Stanley Park (the original Blackpool municipal airport) in October 1941. However, unlike Chester, which handled assembly only, the Blackpool unit was set up (from Weybridge and not Chester) for complete aircraft manufacture from the detail stages. Much bigger than its near-neighbour, this factory had its own machine shops from the outset and its own family of subcontractors.

Altough, like with Chester, a contract was received for the production of 300 Vickers Warwick aircraft, and shortly afterwards cancelled, the entire aircraft output of the Blackpool unit during the Vickers-Armstrongs management period consisted of Wellingtons. The first was completed in July 1940 and the last in October 1945; a total of 3,406 were made (430 of them at Stanley Park).

Similar to Chester, 11,250 prefabricated houses were built at Blackpool between September 1945 and April 1948, before the plant was put on a 'care and maintenance' basis until finally vacated by Vickers on 18 October 1949. It was later turned over to the Hawker Aircraft Co. to resume aircraft production with the Hunter jet fighter from 1952.

From Bomber To Bungalow – A Peace Dividend

During the Second World War, domestic house building perforce came to a standstill in favour of the nationally vital industries, notably aircraft production, needed to support the war effort. But before the end of hostilities, as well as considering the anticipated post-war commercial aircraft requirements and specifications in association with the recommendations of the Brabazon Committee, the British Government also turned its attention to the major problem of housing. Those made homeless by the blitz and the rehabilitation of those coming home from war service were both pressing concerns. There was a predicted major surplus of aircraft skills and materials in the immediate post-war years, until such time as the new civil aircraft designs could be put into

Large-scale production of the de Havilland Dove and Heron light transports, Vampire and Venom jet fighters, and Comet jet airliners in the early 1950s.

production. To complement the Governments plans for a National Health Scheme mooted by the Beveridge Plan of 1944, the MAP commendably organised a competition between aircraft firms for the large-scale production of AIROH (Aircraft Insustry Research On Housing) aluminium prefabricated houses which could be quickly and easily erected on prepared foundations.

The contest for these ingenious utilitarian constructions was won by the Bristol Aeroplane Company and the Vickers Chester factory joined forces with the Bristol-managed Shadow Factory at Weston-Super-Mare, Vickers at Blackpool and Blackburn at Dumbarton. A production order for 11,250 units – 'Prefabs' as they were universally known – was placed at Chester (and a similar number at Blackpool).

This welcome interim work not only provided a most necessary contribution to the revitalisation of post-war Britain but also enabled Vickers to continue the employment of many Chester staff and workpeople in what in today's parlance would be called a 'Peace Dividend'.

The De Havilland Dynasty

The impending completion of all manufacturing work at Chester under the Vickers management in April 1948 meant that the company was scheduled to close and vacate the factory completely soon afterwards with the loss of many jobs.

Fortunately, however, the de Havilland Aircraft Company (DH) at Hatfield, Hertfordshire, had an order book worth some £10million – a considerable sum and business volume at that time. This was principally for its new post-war designs – the Brabazon-specified Dove twin-engined light civil and military transport and the later production Vampire twin-boom jet fighter (after the initial output by English Electric at Preston, Lancashire). The Board of Trade also considered that exports were vital to boost the depressed post-war British economy. These imaginative new DH types were proving widely suitable in helping to fulfil this national mandate in response to the rapid growth of post-war commercial aviation, contrasting with the still uneasy peace that was generating a new round of rearmament by Commonwealth, Allied and friendly nations overseas.

Besides turning out a Dove every working day, the expanding production of the Vampire, plus the continuing output of the Mosquito and the Hornet derivative, the Hatfield plant was

de Havilland Comet jet airliner production in the mid-1950s.

being prepared to take on the development and production of the new Comet jetliner. It had, however, insufficient space and resources to cope with this growth. As had already been realised at the time of the wartime construction of the second large DH factory at Leavesden, noted earlier, for various natural considerations it was not practicable to extend the factory in the Hatfield area. It was therefore decided to lease, if possible, another large factory elsewhere to provide the considerable extra capacity that was urgently needed.

The decision in 1947 to embark upon a large expansion of the company's productive capacity had four main beneficial results. Firstly, the large contracts obtained from overseas purchasers before Britain itself adopted a rearmament policy enabled a strong manufacturing organisation to be built up, ready to expand further when the government did decide to rearm. Secondly, the overseas business speeded the development of the Vampire night fighter and trainer and the Venom variants. Thirdly, these exports consolidated the readiness of several Commonwealth and other countries. Fourthly, they brought in many millions of pounds' worth of nationally important export trade.

Fortuitously, the impendingly surplus Chester factory was considered to be ideal. Moreover, it was larger than Hatfield, was better planned as it had not grown piecemeal, and was the most modern aircraft factory in Europe at that time. Hence, after Wilfred Nixon, DH managing director, and Harry Povey, production director, had gained a good first impression by flying over the Chester site in a Dove, the factory was re-allocated to de Havilland.

As a result, during February 1948, key DH personnel occupied an office in the factory assembly area and began planning the resumption of large-scale aircraft production. As Vickers-Armstrongs was finishing its prefab housing work at one end of the main assembly area, aircraft jigs from Hatfield were being installed at the other. After the formal hand-over ceremony on 1 July 1948, the management of the site was progressively taken over by DH during the rest of that year and the renewed aircraft production work moved ahead quickly.

The first of the many major DH aircraft programmes at Chester was the completion of the end phase of the wooden Mosquito and its wood-and-metal counterpart, the Hornet. The first of the eighty-one Chester-built Mosquitos flew on 30 September 1948, only eight months after the first technicians had arrived from Hatfield, by which time the production of the Hornet and the Vampire (with export orders for Sweden and Switzerland) was well under way. The first of the 149 Hornets flew in March 1949, and Vampire deliveries had started before the end of that year. Total personnel employed reached 652 by December and a recruitment campaign was proving very successful.

The production track of the Hawker Siddeley/British Aerospace 125, the world's first and most successful corporate business jet, and the HS Nimrod main fuselage production line in the late 1960s. The 125 genus continues to be produced at Broughton in airframe component kit form for Raytheon Aircraft Co., USA, who acquired the programme in 1996 after nearly 1000 units had been produced by HSA/BAe since 1962.

The early 1950s saw much diversity of DH designs quantity-built at Chester. In 1950, when the Chipmunk basic pilot trainer aircraft, designed by the De Havilland Company of Canada, was adopted by the Royal Air Force, the bulk of the production in the UK was handed to Chester and 889 of these versatile aircraft were built there during the next seven years. Production of the Dove was transferred to Chester in 1951, to share the shop floor with the Vampire night fighter. The following year, they were joined by the Vampire Trainer and the Venom derivative of the Vampire.

The first of the 424 Chester-built Vampire Trainers and 834 Venoms were completed in 1952. Thus in only its first four years under DH management the Chester factory had produced aircraft of eight distinct types – seven of which were being built simultaneously (Mosquito production having ceased in November 1950).

In 1953, DH employed 12 test pilots to cope with the large numbers of Vampires, Venoms, Chipmunks, Doves and Herons which were being produced to meet the dual requirements of the post- Korean war re-armament programme and rapid growth of civil air transport.

The Comet jet airliner, first flown at Hatfield on 27 July 1949, showed early market promise on both sides of the Atlantic. This success led to de Havilland and Short Brothers and Harland establishing an agreement whereby a second Comet production line would be set up at Belfast in Northern Ireland. On 31 October 1952, DH announced that some component manufacture for the Comet had already started at Chester and that it intended to lay down a third Comet line there. All three lines would make the Comet 2 and were expected to make the changeover to the Comet 3 in 1956-1957. The long-term plan was based on Hatfield being the development and pre-production base, with quantity production shared between the three factories.

Work on the Comet began at Chester, initially sharing with Hatfield the production of the Comet 2 for British Overseas Airways Corporation (BOAC), and notably including extensive use of the metal-bonding process. However, when the first of the twenty-five Comet 2s in the initial production plan was nearing completion in 1954, disaster struck the programme. There had been three aircraft losses in mysterious circumstances in BOAC service and production of the first twelve

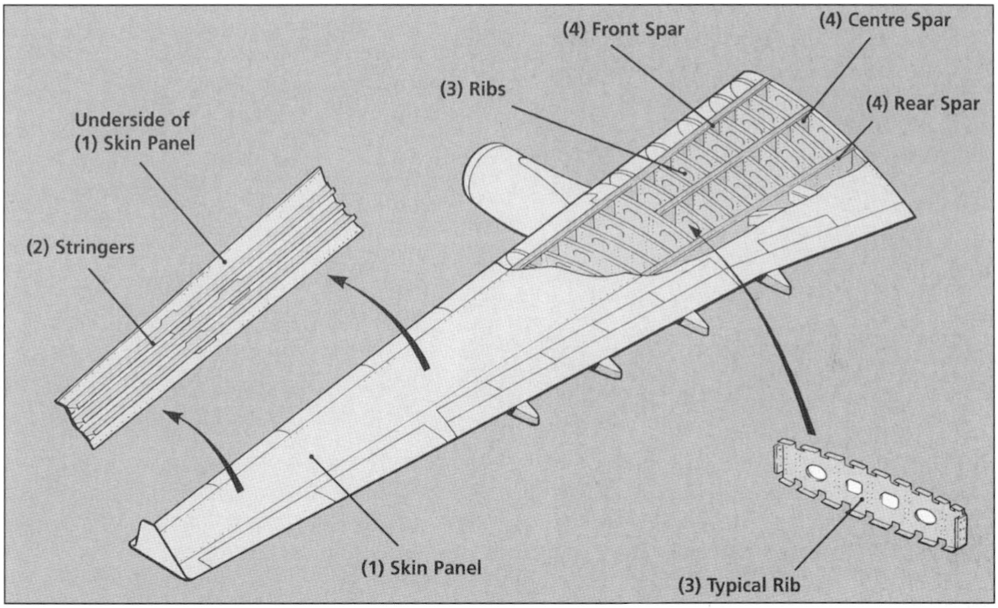
(4) Front Spar
(4) Centre Spar
(3) Ribs
(4) Rear Spar
Underside of
(1) Skin Panel
(2) Stringers
(1) Skin Panel
(3) Typical Rib

The principal elemental components in the construction of a typical Airbus wing torsion box/integral fuel tank structure, of which over 2,600 sets (pairs) have been built at Broughton – the exclusive manufacturing centre for Airbus wings since the start of the programme in 1971.

machines in progress there had to be abruptly suspended. Components which had been completed at the similarly moribund Belfast line were later also shipped to Chester. Restarting production in 1955 by reworking the Mk.2 airframes with strengthened fuselages, principally to serve as transports with the Royal Air Force, meant that the first (and only) Comet 2 to be completed and flown at Chester was XK716. It was made for the RAF and was delivered to No.216 Squadron on 7 May 1957 and named 'Cephus'. Chester's Comet 2 up-date programme was completed in 1957. The following year, when the full cause of the accidents had been ascertained by the Royal Aircraft Establishment (RAE) at Farnborough, and the remedy devised, a Comet 4 production line was established at Chester to augment the Hatfield line. The first of this modified and extended version from the Chester line, G-APDE, was delivered to BOAC in 1959 and production of the Comet 4 continued until 1964. Overall, Chester had built forty-two complete Comet airframes of the total Hatfield/Chester output of 117 units (including test examples).

1953 saw the start of production batches of the Dove's larger four-engined derivative, the Heron, with 244 and 140, respectively, being built when production of both types was completed in 1967. Significantly, the Heron output included a fleet of four VIP aircraft built for HM the Queen's Flight.

Achieving the huge production build-up and export performance at Chester, handling the main flow of DH civil and military transport and combat aircraft during the 1950s and 60s, was not without very real difficulties, especially in the recruitment of suitable labour in the largely agrarian locality. This required extensive training so, in early 1949, a branch of the famed de Havilland Technical School (originally formed at Hatfield in 1928) was opened to support the prolific apprentice scheme, and another for tradesmen, within the factory. Workers were attracted from as far afield as Belfast and Glasgow. Long-distance bus services were operated for daily commuters from centres as far distant as Liverpool, Warrington and Crewe. The local authorities of Chester, Hawarden and Mold also greatly assisted with new housing.

Tooling and equipping was a constant priority, especially with the introduction of the vastly more complicated and larger-scale Comet jetliner, with its much higher man-hour content and new production processes.

The large-scale machining of Airbus wing skin panels. This process has consistently involved some of the largest machine tools anywhere in the world since the inauguration of the programme in 1971.

Under The Hawker Siddeley Banner

The government enforced the consolidation of the UK aircraft industry in 1960 into two main airframe groups – British Aircraft Corporation (BAC) and the Hawker Siddeley Group (HSG). This resulted in the, hitherto independent, de Havilland Aircraft Company being taken into the latter group. This was the second of the many parental organisational changes to embrace the Chester unit. It was, however, a seamless process at the working level and the pioneering DH name continued in its new designation as part of the Hawker Siddeley de Havilland Division in conjunction with Hatfield.

Assembly of the DH Canada Beaver light transport aircraft, chosen by the British Army Air Corps, began at Chester in 1961, continuing into 1962 and resuming with a repeat order in 1967; a total of forty-six aircraft were eventually built. Production of the Sea Vixen naval all-weather fighter began in 1962 and continued until 1966 with the completion of thirty aircraft for the Royal Navy.

On 13 August 1962, the first flight took place at Hatfield of the distinctive DH125 corporate business jet that was soon to become a veritable lifesaver for Chester. Intended as a jet-powered successor to the Dove, the concept of this diminutive transport aircraft had been first revealed publicly in 1961 as: 'the World's first jet aircraft specifically designed as a corporate communications aircraft' and 'a jet airliner in miniature'. Because Hatfield was becoming heavily involved with the development of the much bigger Trident airliner, Chester thus became the exclusive production source for the 125 and the first batch was laid down there at the beginning of 1962.

The Hawker Siddeley Group of engineering and aviation companies was restructured on 1 July 1963, and its aviation interests incorporated in a new subsidiary, Hawker Siddeley Aviation (HSA). Chester was brought within the HSA de Havilland Division, and all but the longest established DH products were re-designated with the 'HS' prefix.

But with the end of the Comet line in sight, rumours were rife that both the Broughton factory and the Hawarden airfield might have to close. However, it was the encouraging build-up of orders for the 125 that doubtless saved the day and kept a large part of the factory busy through the 1960s.

It was then that the long-famous de Havilland company name disappeared completely, passing into history on 1 April 1965 after seventeen years currency at the Chester plant (and

The first set of Hawker Siddeley Aviation Chester-built wings for the Airbus A300 airliner being loaded aboard a Super Guppy transport aircraft at Manchester Airport on 23 November 1971. They were bound for (what was then) the VFW factory at Bremen, Germany, for equipping prior to onward flight to the Aerospatiale factory at Toulouse, Southwest France, for final assembly.

forty-five years altogether since established by its pioneer signatory, Sir Geoffrey de Havilland, on 25 September 1920). HSA abandoned all of the original company names of its constituent divisions in favour of geographical designations. Paradoxically, however, the bonus was that the Chester name came forward formally for the first time when it was embodied in the title of the new entity, the HSA Hatfield-Chester Division.

Between 1965 and 1967, two Chester-built speculative Comet 4Cs were re-allocated to become development prototypes for the Nimrod maritime reconnaissance and anti-submarine aircraft, to replace the Avro Shackleton of RAF Coastal Command (a large batch of which had earlier been overhauled and up-graded at Chester). An assembly line of this, the world's first pure jet maritime aircraft, was started in 1966 with the production of the main fuselage and other components. These were then transported to the HSA Manchester Division factory at Woodford for completion with the lower fuselage, final assembly and delivery. The Chester involvement in this programme ceased in 1970 after forty-one airframe sets had been built there.

In a significant step by HSA, following a number of exchanges of new airliner design concepts with other major European aircraft manufacturers in the mid-1960s, the company became the UK industrial partner when a tripartite Memorandum of Understanding (MOU) was signed by the British, French and German governments. It was signed on 26 September 1967, and set out the plans for the design of a new-generation wide-body, twin-aisle, twin-jet airliner. The agreement was to provide the expertise required for the advanced aerodynamic design and manufacture of the wing which HS had acquired with its Trident programme and for which there was no equivalent within the two continental partner industries. HS kept its faith in the programme even after the British Government withdrew its support in March 1969. When the Airbus Industrie (AI) airliner programme was launched on 29 May 1969 by the French and German governments, HSA continued to participate in a private sub-contract role, taking responsibility for the design and

The 100th Airbus A300 wing leaving the assembly jig in June 1979. The 2,000th set was delivered just short of ten years later in February 1999.

manufacture of the complete main wing structure for the definitive Airbus A300. HSA then sustained its involvement in the programme when AI was formally constituted in December 1970.

Significantly, this action resulted in a new and enduring dimension of scale, capability and output for Chester from 1971 with the start of construction work for this major new multinational civil aircraft venture. In turn, it lead to the installation of some of the largest and most advanced structural machine tools and manufacturing jigs anywhere in the World.

The first completed A300 wing set was despatched on 23 November 1971. The transportation of the huge structures was initially by road on custom-designed trailers to either Manchester or Liverpool airports, where they were loaded into a 'Super Guppy' specialist transport aircraft. They were then delivered to the VFW plant at Bremen in Germany for equipping, prior to being flown on to the AI final assembly centre at Aerospatiale at Toulouse. This process thus ensured that no structural component would ever be out of production working for more than forty-eight hours.

British Aerospace and Airbus Wings the World Over

Six years on from the start of the Airbus venture at Chester, the plant was re-designated again as a result of the nationalisation edict of the British aircraft industry by the then Labour Government. In this process, the Hawker Siddeley Group's aviation and guided weapons interests ceded into the newly established British Aerospace (BAe). After 'vesting day' for the new company on 29 April 1977, the formal constitution came into being on 1 January 1978. The former Hawker Siddeley Aviation (HSA) company became, in turn, a member of the BAe Aircraft Group with the Chester unit and name continuing within the BAe Hatfield-Chester Division. Agreement at the industrial level between the AI consortium partners was initialled on 18 August 1978 and ratified by their respective governments on 27 October so that BAe joined AI on 1 January 1979 as a full decision-making partner with a 20 per cent shareholding.

As well as the expanding Airbus airliner customer base, sales of the (now further redesignated) BAe125 were likewise continuing steadily and the sale of the 500th aircraft was announced in

The completely modernised, extended and equipped and highly productive BAe Systems Airbus UK Broughton factory in 2000 (August).

October 1980, after a record-breaking sales year in 1979 during which fifty-five aircraft were sold. The 100th set of Airbus wings was also delivered from Chester to Bremen in June 1979.

Further corporate organisational change came on 1 January 1981 with the formation of British Aerospace PLC, following its flotation on the Stock Market and substantial re-privatisation (with the return to full privatisation completed on 10 May 1985). It was also in 1981 that BAe purchased the freehold of the hitherto leased Chester factory site and airfield from the Ministry of Defence. The first wing set for the A310, the second member of the AI family, left the jig at Chester on 7 April 1981 and after equipping at Filton was despatched to Bremen for completion on 17 May 1981.

Yet another change came in early 1989, following the formation of British Aerospace (Commercial Aircraft) Ltd on January 1 and the establishment of three autonomous business divisions of that company a month later; the Chester site became part of the new BAe Airbus Ltd – in conjunction with the BAe Filton (Bristol) site. However, this change also meant the loss of the Chester headline name after twenty-four years use – seven years longer than the site had operated under the de Havilland name banner. At this point, Chester also became the output centre of BAe Corporate Jets Ltd, the company responsible for the continuing BAe 125 business. Renamed Corporate Jets Ltd in May 1992, the following year the company was sold by BAe to the Raytheon Aircraft Company (RAC) of the USA – which continues to apply the 'Hawker' prefix to its latest derivative models. Airframe component kits continue to be made at Broughton under a subcontract manufacturing agreement for assembly by RAC in the USA.

Airbus wing production built up at an almost exponential rate through the 1980s. The first wing set for the new single-aisle addition to the AI family, the 'reference standard' A320 regional single-aisle twin-jet, was despatched from Chester to BAe Filton, Bristol, on 28 November 1985, to a new A320 Wing Equipping Centre there. It then went on (initially by road and sea via Plymouth, and later direct by Super Guppy aircraft) to Toulouse for final assembly. This sequence continued until July 1993 with the completion of the 453rd A320 wing set, whence the equipping of these wings was reverted to Chester.

The highlights of 1989 were the commissioning on 1 August of the new £5million 'link' building (between the original main assembly area and the former flight shed) and delivery of the 600th

The first 98ft/30m long wing structure for the Airbus A340 – the first of Airbus Industrie's very long-range airliners – ceremoniously leaving the assembly jig in late June 1990.

Airbus wing set in September. The first wing set for the A340 wide-body, twin-aisle four-jet intercontinental jetliner was delivered in August 1990, the 250th A320 wing set (from Filton) in July 1991, and the first wing set for the twin-engined wide-body A330 counterpart of the A340 in September. This coincided with the opening of a new £3.2million Airbus Wing Despatch Centre.

During the 1990s, the ageing Super Guppy aircraft were replaced by a fleet of the specially-adapted Airbus A300-600ST 'Super Transporter' aircraft, incorporating Chester-made wings – and named 'Beluga' because of the resemblance of a giant whale of that name. Following appropriate improvements to the main runway in 1996, these extraordinary aircraft transported all Airbus wings direct from the Hawarden aerodrome instead of Manchester airport after road transfer. Those for the A300/A310/A330/A340 wide-body lineage are first flown to Bremen for the equipping before proceeding to Toulouse for final assembly. Those for the A319/A320/A321 regional twin-jet family are flown, pre-equipped, direct to Toulouse (A320) and to Hamburg (A319 and A321) for final assembly in the complete aircraft.

The 2,000th Airbus wing set was delivered in February 1999, the year in which Airbus Industrie commendably won an equal market share of the intensely-competitive global market for mainstream jetliners against its formidable American competitor, the long-established Boeing Company, and to the considerable benefit of the Broughton factory. Appropriately, British Aerospace Airbus also won a Queen's Award for Export Achievement that year.

BAE SYSTEMS and Airbus UK Into The New Millennium

The Filton and Broughton (now the exclusive North Wales site name) teaming was operating from 31 March 2000 as the manufacturing centre of Airbus UK Ltd, the wholly-owned subsidiary of BAE SYSTEMS, which had become effective from 30 November 1999 with the merger integration of the former British Aerospace (BAe) and GEC-Marconi Electronic Systems (MES) companies. The formation of Airbus UK Ltd was in anticipation of the transformation of the now anachronistic French Groupment d' Intérêt Econonico (GIE), legal format of the four-nation Airbus Industrie organisation (UK, France, Germany and Spain), into a fully commercially-accountable Single

The BAE SYSTEMS name banner erected on the main office block at Broughton in 2000 following the formation of the new company on 30 November 1999. For six decades, Broughton was the only aircraft manufacturing plant in Wales. It has since been complemented by the formation of Airbus UK on 31 March 2000 and the announcement of the new-concept Airbus Integrated Company (AIC) on 23 June 2000, which became operationally effective from 1 January 2001.

Corporate Entity (SCE). It was in the form of a French S.A.S. (Société par Actions Simplifiée) or simplified stock company, in which Airbus UK represents BAE SYSTEMS 20 per cent share holding. The newly-constituted European Aeronautic Defence and Space Company (EADS), which was formed on 10 July 2000 by the merger of the French Aerospatiale Matra, German DaimlerChrysler Aerospace and Spanish CASA , now has the remaining 80 per cent shareholding in the new Airbus company. In this simplified and unified organisational arrangement of Airbus interests, the Broughton wing manufacturing facility continues to operate integrally with the Airbus UK management, design and manufacturing centre at Filton, Bristol.

At present, the 450-acre Broughton factory/Hawarden aerodrome site employs a workforce of around 4,300 people from Northeast Wales, Chester, the Wirral and Liverpool (more than three-quarters of its peak wartime personnel complement). It injects £3.5million a week in wages, goods and services into the economy within a thirty-mile radius of the plant.

Major capital investment in new factory equipping and extensions were implemented during 1999 for the production of the wings for the latest four-engined 335-450 seat A340-500/600 extended-range developments. The first sets of wings (with 100ft long skin panels) for these two important new committed derivatives, was delivered in April 2000. Those for the 108-seat single-aisle A318, the smallest in the entire Airbus range, are being completed in August 2001. With the significant milestone of the delivery of the 2,500th Airbus wing set (i.e. 5,000 individual wing structures) on 12 September 2000, high-rate production continues at Broughton for every member of the Airbus airliner dynasty.

The completely new Airbus A380, formerly known as A3XX, 480-650 seat double-deck commercial airliner project – the World's largest – is the next challenge for the newly-reconstituted Airbus enterprise. Consequently, a considerable increase in employment, work throughput and export earnings is anticipated for the Broughton site in the near future. Broughton received a £530million repayable UK Government loan investment committed in May 2000 to the launch of the A380, which was complemented by BAE SYSTEMS' own planned investment of £230m in the site over the next five years. The Welsh National Assembly First Secretary announced on 24 September 2000 a Welsh Development Agency financial package grant totalling £19.5million to support training and skills development, which will also enable continuing capital and skills investment at Broughton for the manufacture of Airbus wings and Raytheon Corporate Jets. Significantly, this grant also confirmed that the

The next-generation Airbus A380 outsize double-deck airliner – the launch of which was announced on 19 December 2000. It augurs a new physical scale and level of technology in transport aircraft wing design and manufacture for Airbus UK in the new Millennium.

manufacture of the A380 wing structure has been secured for the North Wales site, thereby preserving the 3,000 existing jobs and creating a minimum of 1,700 new ones.

The Broughton six-decade heritage, reputation and record will bequeath to the A380 programme a correspondingly exceptional and enhanced capability and value. This portends yet further and greater diversity, business volume and success for Broughton as a globally-acclaimed wing centre of excellence for the foreseeable future.

Left: Sir Robert McLean, Chairman of Vickers (Aviation) Ltd, who approached the British Government in 1936 with the need for another Wellington production factory to augment the output at the Vickers parent plant at Weybridge. Centre: Trevor Westbrook, Vickers General Manager – Weybridge, Chester and Blackpool, who had transferred to Weybridge from Vickers-Supermarine at Southampton at the end of 1936 after having been the driving force in getting the all-metal Spitfire fighter into quantity production. Right: John Corby, de Havilland's first General Manager, Chester: 1949 – 1958.

'Wellington Warriors': The Vickers-Chester management with a group of visiting Royal Air Force Wellingtom aircrew in 1940. Centre row, third from left: Ronnie Yapp, Commercial Manager (later Managing Director of Vickers Ltd). Fifth from left: Gordon Montgomery, Works Manager. Front row, first left: Maurice Hare, Chief Test Pilot. Centre: Bernard Duncan, Superintendent.

Management Personalities

After the initial short period at Chester, during which Gordon Montgomery was seconded from Weybridge as Assembly Manager (responsible for the first Chester-built Wellington), the General Manager of the Vickers-Armstrongs Weybridge, Chester and Blackpool units was the redoubtable Trevor Westbrook. He left the company on 30 March 1940 to join the Ministry of Aircraft Production (MAP), Lord Beaverbrook's wartime team of aircraft production expeditors. Under Westbrook, the Works Manager at the Chester factory was E.R. 'Rex' Talamo (brought in from elsewhere in the Vickers Group), who went on to become the Superintendent after Westbrook vacated his office. Talamo was then replaced as Works Manager by Gordon Montgomery, who sadly died in 1941. Talamo was later transferred to the Vickers Spitfire Shadow Factory at Castle Bromwich, Birmingham, and was succeeded at Chester on 1 May 1940 by Bernard Duncan (who had previously been Experimental Manager at Weybridge and was succeeded there by the subsequently knighted George Edwards). Duncan then retained the office of Superintendent until 30 June 1948 when all Vickers activities ceased at the Chester factory.

The de Havilland era at Chester was inaugurated by a small team from Hatfield, with Clem Pike, personal representative of Wilfred Nixon (DH Managing Director), in charge of preliminary arrangements. The first permanent appointments were: John Corby, General Manager; James Mackenzie, Production Manager; Sydney Statham, Factory Manager; Arthur Turner, Production Engineer; and David Brown, Chief Inspector. Successive 'directors-in-charge' during the subsequent de Havilland, Hawker Siddeley, British Aerospace, BAE SYSTEMS and now Airbus UK tenures (albeit with varying titles) were/are:

Arthur Turner, Production Divisional Manager: 1959-1961 (member of the original team which re-opened Chester for DH in 1948).

Jack Garston, Production Manager: 1961; Works Manager: 1962-69; Executive Director and Works Manager: 1970-71; Executive Director and General Manager: 1972-77.

Percy Edwards, Director and General Manager: 1977-79.

Arthur Rowland, Director and General Manager: 1980-84.

John Gillbanks, Divisional Director and General Manager: 1984-88.

Shaun Dyke, Director and General Manager: 1989-1995.

David Waring, Director of Manufacturing and Product Executive Hawker Jet: 1995-97.

Bill Travis, Director of Manufacturing: 1997-99.

Brian Fleet, Director, Manufacturing, Airbus UK.

One
Building A Bomber Birthplace

Construction of a large-scale aircraft factory and supporting airfield on a 'green-field' site is rare. The massive construction task for the Chester Shadow Factory was compounded by urgency, and the fact that the site was poorly drained and had a significant sloping aspect requiring major levelling.

The civil engineering contract for the factory was awarded to Sir William Arrol and Son of Clydeside. Ground-breaking began on 14 December 1938 and the entire complex was completed in nine months, long before the advent of modern-day earth-moving equipment and tower cranes. It was fully operational by the outbreak of the Second World War in September 1939. The factory was the largest covered building in Europe with a total floor area of 1.5million sq.ft. Large elements of the fabric of the original construction are still in use today.

The virgin construction site, indicating the rural nature of the area and the preparation necessary prior to construction. The orientation is shown by the L-shaped spur from (what is today) the A5104 Chester Road (lower centre) which became the main entrance to the complex, leading to the main office block.

The construction site had initially appeared level, but there was actually a fall of about seven feet from south-west to north-east. This necessitated the excavation of four feet at the former end and the filling up of three feet at the latter end. Trenches had to be cut for the new drainage system to direct existing gullies away from the factory site. Despite excessive winter flooding, small temporary offices for Arrols and Vickers-Armstrongs were erected and the first strip of top soil removed before Christmas 1938. The railway spur line sidings, adjoining the London, Midland & Scottish (LMS) line, were delivering the first drainage pipes. However, by January 1939 the exceptional winter rainfall meant that the site became a sea of mud and a network of jubilee tracks had to be laid (centre) to enable the completion of the earth-moving operation.

By early March 1939, the first vertical steelwork was being erected. By the end of the month, the steel erection was well advanced and a start had been made with the concrete columns for the offices. A temporary rail track (foreground) encircled the building work to enable materials to be transported direct to the point of use. Poles carrying temporary lighting for night work were erected around the whole site.

Steel erection of the 'low' part of the main factory building nearing completion, pending attachment of the side-wall panels.

Steel erection of the 'high' main assembly area and 'Northern Light' roof structure. The large trestles under the transverse roof support beams were temporary constructional aids prior to this area becoming a completely unsupported roof span. The whole factory was later traversed by a unique and powerful overhead crane system attached to these beams, which could lift up to ten tons at any point using two five-ton cranes. Complete wings and power-plants could be swiftly lifted into position over the wheeled-trolley transported Wellington fuselage line. Twenty years later, this crane system would enable Comet jetliner wings and fuselages to be moved in order to be united within minutes and, indeed, the same goes for the movement of Airbus wings today.

The main office block and forecourt, fronting the first bay of the main factory construction, was finished and occupied in mid-September 1939 (prior to the later addition of a second floor and hipped roof). It has the Vickers-Armstrongs Ltd signage prominently displayed above, and on, the entrance reception doorway. Many of the original Vickers-Armstrongs 'V-A'-identified lamp standards (bottom right), that were to line the site approach road, survived until recently when they had to be removed because of 'concrete fatigue'.

Although the Chester factory location was believed remote from possible enemy air attack, camouflage was applied early to the front of the main office block. This was made from the hessian used for protecting the setting concrete during the wet winter weather.

The main factory, largely completed externally, and camouflage – painted in mid-November 1939. The adjoining land was still very muddy and relying on horse-drawn transport in many areas.

The north-west corner of the engine-running and flight shed near completion in mid-November 1939. Sliding doors are being fitted, as is a section of the fencing (bottom left), which ultimately enclosed the whole factory site.

By mid-November 1939, the brickwork and roofing of the 1,500-seat staff canteen was well advanced, with the works transport vehicle park being accommodated underneath. Though much modified and updated, this building continues to serve as the staff restaurant and site transport centre today.

The interior of the staff canteen main hall with self-service on either side. The entertainment stage facility at the far end was frequently featured in the famous national morale-boosting BBC lunchtime radio programme, Workers Playtime, announcing, as wartime security demanded, that is was 'coming to you from an aircraft factory somewhere in the North of England.'

The Hawarden Aerodrome

The pre-existence of flying activity on the original Hawarden Estate was not only a significant factor in the Air Ministry requisitioning the site for the flight test and delivery of the output of the adjoining Shadow Factory but also meant that it naturally lent the Hawarden name to the whole operation in the minds of many. It was located on fertile but poorly drained agricultural land reclaimed from the River Dee when a five-mile section between Shotton and Chester was canalized in the eighteenth century. A 100-acre stretch of this area had been used from 1935-1937 as a relief landing ground (RLG) for RAF Sealand, three miles to the north-west.

During construction of the factory from mid-December 1938, the land was compulsorily purchased and it comprised of about a third of the area of the present site, on a strip of land lying south to north. The civil engineering contractor was Bernard Sunley and Co. Ltd. Despite considerable land in-filling and the installation of sophisticated pipe-work, the absence of natural drainage had serious consequences over successive winters, not only hampering construction of the factory but also wartime operation of the airfield. Most of the grubbing of trees and hedgerows was only made possible by using pairs of steam ploughing traction engines located 100 yards apart and hauling large drag buckets and multi-share ploughs between them on heavy wire ropes.

Enlarged during 1939, and taking advantage of the high local unemployment, Hawarden Rural District Council co-operated in building new housing estates for the airfield and factory workers, the first at Broad Oak, Broughton, with others following at East Saltney and Ewloe. After the first hangars had been erected in early 1940, the airfield suffered its first and only air raid on the night of 14 November 1940, the night of the infamous air attack on Coventry, damaging twenty-eight aircraft in two hangars.

Although never used for operational flying, the main wartime occupation of the airfield was by the major RAF Maintenance Unit (No.48 MU) formed on 6 March 1940 (ultimately disbanded on 1 July 1957). The unit also occupied most of the surrounding fields with aircraft storage and all the hangars (most of which still survive). No.7 Operational Training Unit was also formed there on 15 June 1940 (becoming No.57 OTU) for fighter and fighter-reconnaissance pilots and Air Transport Auxiliary (ATA) Ferry Pilots Pools (FPP) for the delivery of new aircraft, occupying hangars near the control tower. Fighter training was perfected using a early form of flight simulator, locally devised, known as the 'Hawarden Trainer'.

The RAF's main function at Hawarden during the immediate post-war years was the scrapping of many hundreds of surplus wartime aircraft, although it did continue to prepare new ones for service use on a much reduced scale. When de Havilland took over the factory in 1948, the airfield was initially used for flight-testing by arrangement with the Air Ministry – three jointly-sponsored Battle of Britain displays were also held there in September 1951, 1954 and 1956. Continuing in dual use by the RAF and DH, the military station accommodated No.173 Squadron to ferry aircraft in the UK, until its disbandment on 2 September 1957. RAF Hawarden station was ultimately closed on 31 March 1959, due to major cuts in defence expenditure, after which de Havilland took over the complete site, inheriting the vacated hangars and buildings, thereby ending Hawarden's twenty-year association with the RAF (although a gliding school survived until 1963).

In 1960, DH attempted to generate revenue by starting commercial services, at what was officially known as 'Hawarden Airport', which continued spasmodically until June 1979. Much further development of the aerodrome has taken place over the last twenty years – most notably, the strengthening and widening of the ends of the main 6,000ft/2,043m runway and the provision of 'bat-handle' ground-turning areas for the specialised wing transport aircraft in 1996. This has enabled all Airbus wing structures to be despatched direct from Hawarden, instead of either Manchester or Liverpool airports.

Overhead the Broughton/Hawarden factory and aerodrome complex in July 1947, two years after the end of the exhausting Second World War, transformed to build prefabricated houses as a 'peace dividend', with many war-surplus aircraft still parked awaiting disposal. The remarkable wartime production record by Vickers-Armstrongs of 5,786 large bomber aircraft (and the repair of many more) was clear testimony not only to the intelligent design, construction and operation of the complex as an autonomous aircraft manufacturing plant, but also to the success of the British Government's imaginative 'Shadow Factory Scheme.'

The geographical orientation of the factory and main 05-23 runway running parallel to it, is approximately south-west to north-east from the main entrance end of the site (the orientation of this photograph being from South (top) to North (bottom) along the axis of the secondary runway).

Central Registry of Air Photography for Wales.

Two

The Vickers Bastion Of Big Bomber Production

The designation of the Chester shadow factory for large-scale production of the Vickers Wellington from scratch, with a largely untrained workforce, remote from the main Vickers plant at Weybridge, posed a major challenge in addition to having to cope with the complex geodetic ('basketweave') airframe structure. Essentially operating as a 'parent works', the manufacture of all components was subcontracted, including nearby dispersal sites and a second assembly line at Cranage aerodrome from 1941. By the end of 1940, 490 Wellingtons had been built, followed by 897 in 1941, 1,356 in 1942, 1,356 in 1943, 1,217 in 1944, and 224 in 1945 (up to September when production ceased after the completion of the 5,540th aircraft). This total comprised three Mk.I; seventeen Mk. IA; 1,583 Mk. IC; 737 Mk. III; 220 Mk. IV; 2,434 Mk.X; eight tropicalized Mk. XII; and 538 Mk.XIV aircraft. 235 Avro Lancaster heavy bombers and eleven Lincolns were also built between June 1944 and September 1945. The end of the war meant that all aircraft manufactured at this Vickers bastion of big bomber production ceased after a grand total of 5,786 aircraft had been produced in exactly six hectic years – an overall average of more than 1,150 a year.

The Vickers-Armstrongs Ltd company title together with twin stone-carved interlaced 'V-A' logos prominently featured above the main entrance and in the glass doorway panels of the Chester site management and administration block.

Because of the delay in the ground levelling due to the quagmire and the consequently difficult surface tracking conditions, the contractors' promise to deliver the south-west corner of the main factory site by the third week of February 1939 could not be kept. A Bellman-type hangar was therefore borrowed from the Air Ministry and erected on the edge of the site near the Glynne Arms public house in order to begin assembly of the first aircraft while simultaneously providing on-the-job training of local workpeople.

The first pair of Wellington fuselage assembly jigs and main structural frames in place in the temporary Bellman hangar with geodetic panels available to start construction of the first Chester-built aircraft in April 1939.

The first Wellington fuselage assembly jigs installed in the south-west corner of the new factory, the first area to be completed and occupied in mid-June 1939. The partially completed assembly of the first aircraft was then transferred from the Bellman hangar to this area in mid-July alongside the first batch of completed skeletal geodetic fuselage structures.

The first Chester-built Wellington, a Mk.1 L7770, assembled in the four months between the start of work on 3 April 1939 and first flight on 2 August 1939, before being delivered to the Royal Air Force just at the outbreak of war on 3 September 1939. This was a remarkable achievement under the guiding hand of Gordon Montgomery, Assembly Manager, with a small team and parts from the Vickers-Weybridge factory, while working alongside the construction of the main factory and simultaneously training the nucleus of staff for its occupation. Sir Robert McLean, the Chairman of Vickers (Aviation) Ltd, (dark suit) and Montgomery (with hat and raincoat) are observing immediately in front of the nose turret.

Take-off of Wellington L7770 on its first flight on 2 August 1939. Considerable difficulty had been experienced in moving the aircraft out of the unfinished factory and onto the newly constructed Hawarden aerodrome. This was because the drainage system had not proceeded as quickly as hoped and large areas were under water, leaving only one end available for flying. Consequently, the aircraft had to be flown to Weybridge for full flight clearance. By late 1940, only having laid a single short (700-yd) runway (still used today as a taxiway and known as the 'Vickers Track'), the contractors were hard pressed to build all the facilities that were so badly needed that wet winter.

The Vickers patented drawbench machine, incorporating an elaborate cam and roller control mechanism for forming and shaping the multitude of individual geodesic channel members embodied in the Wellington airframe that were hand-sawn to length. This ingenious machine can be fairly regarded as a forerunner of the modern generation of computer-controlled machine tools. Three firms in the Midlands undertook all the geodetic rolling mill work for the geodetic construction of the Wellington at Chester.

The Geodetic Structural Concept

The unusual geodetic ('basketweave') structural principle deployed so effectively in the Vickers Wellington was the brainchild of the legendary Barnes Wallis and originated in the method that he devised for the retention of the gas-bags of the Vickers R100 airship of 1929. Because of the difficulty in estimating the structural loads, the Airworthiness and Airship Panel would not allow the retention wires to be fixed to the longitudinal beams. Contriving a spirally-wound retaining wire mesh attached to a secondary structure, Wallis sought the guidance of Professor Filon, Professor of Mathematics at University College, London, who advised that the lines required were 'geodesics' and provided the analytical equations.

Geodesy, from which this concept was derived, is a branch of mathematics dealing with the measurement of the shape and surface of the Earth, wherein a 'geodetic line' or 'geodesic' is defined as the shortest distance between two points on a curved surface and known in global navigation as a 'great circle' route. When Wallis joined Rex Pierson, Vickers Chief Designer at Weybridge, as Chief Structural Designer in 1930, he reasoned that this concept could have an analogous meaning in aircraft design by enabling the stresses to be carried by the shortest route in an ideal load-balance and fail-safe combination. It could also replace normal primary and secondary members with a self-stabilising system of main members only, doing the work of the shell of a conventional monocoque, without the need for internal load-carrying structure. Near-ideal streamlined external shapes could thus be adopted, together with maximum unobstructed internal space and high strength and stiffness. Stressed-skinning could also be dispensed with, and well-practised, minimum-weight, fabric covering still be used.

By helical-winding the load-bearing structural members in opposite directions along the length of a substantially single-curvature tubular fuselage, and joining them at each crossover, one set of 'geodetic bars' was in tension while the other was in compression. The resulting curved diagonal lattice, stabilised by quartile tubular longerons, thus lay along the lines in which the principal in-flight forces acted and absorbed all loads by stress equalisation in sustaining the bending, shear and torsional loading generated by the aerodynamic forces on the airframe. As applied to a wing, tailplane or fin, this structural form also afforded exceptional torsional stiffness, allowing a high aspect ratio (for long range), without the tendency to flutter or aileron reversal as with a more conventional design.

First applied to the Vickers G.4/31 biplane prototype of 1934 and its production derivative, the Vickers Wellesley long-range monoplane bomber of 1935, this ingenious, if complex, structural system was fully exploited in the large-scale high-rate production of the Wellington. Axiomatically, completely new design and manufacturing processes were necessary to translate the Wallis concept into practical reality. These imperatives came from the ingenuity and drive of three other Vickers-Weybridge stalwarts. Basil Stephenson, in the design office, was responsible for geometric/spatial definition of the nodal co-ordinates of the lattice structure in relation to the airframe external shape. This permitted the mathematical definition of the unwrapped spiral members in a series of chordal lengths and intersecting angles to be expressed in tabular form. Although Professor A. J. Sutton Pippard of London University provided Wallis with a method of stress analysis, this was not used and Vickers developed its own. Jack East of the tool room then devised an ingenious strip-rolling machine, using an intricate arrangement of adjustable sets of cams and rollers, to form and shape the multitude of individual channel-section members. The whole concept was then brought into mass production simultaneously in the three Wellington factories – Weybridge, Chester and Blackpool – by the redoubtable Trevor Westbrook. In active service, the geodetic-type construction of the Wellington brought the added bonus of its ability to withstand major battle damage and still enable flight crews to 'get home' safely.

Boiler-plate sheets drilled and pegged to check the profiles of the geodesic members after forming.

Assembly of the geodesic members in diagonal lattice pattern.

Pre-assembled fuselage geodetic panel stores supplied from subcontractors. The Wellington structural skeleton consisted of curved channel-section members ('geodetic bars') laid in opposite directions. Those running in one direction were cut halfway through on the inside of the curve and those running in the other direction were cut on the outer surface – and inter-connected with specially designed machined 'butterfly' and 'wishbone' fittings, rivets and bolts. The bars were then joined to the fuselage stabilzing tubular longerons (or wing spar) by shear cleat and gusset plate attachments.

The first semblance of a Wellington assembly line in the main factory in mid-September 1939, with the second machine nearing completion at the far end of the shop, followed by two lines of fuselages and then a pair of wings. Bottom right: Further jig support structures awaiting assembly.

Fully-assembled Wellington geodetic fuselages being fitted with overlaying bolted longitudinal wooden 'fabric rails' for the attachment of the Irish Linen fabric covering. The geodetic panels were broken at the longerons for ease of construction, but as the lengths of these longerons were short, the resulting bending moment in them was very small. They took the tension and compression loads due to bending, and the geodetic bars the shear and torsion, while also supporting the longerons against buckling.

Wellington fuselage at an advanced stage of the fabric skin covering. Attachment of this covering to the geodetic structural members was achieved by capping strips screwed to the wooden fabric rails to prevent chafing against the structure and edge-sewn, eight stitches to the inch. The screws were first bench-fitted to the rails by a near-blind operator who drove around six thousand screws a day. Each fuselage then incorporated eight thousand small bolts in the fitting of the rails to the airframe.

Background: Fabric covering, stitching and dope stretching of Wellington wing structures. For the wings and similar surfaces the fabric was looped around a securing wire reinforcement and stitched through small holes in the outside flange of the geodetic bars, with the stitches being securely knotted on the inside. Foreground: Application of identification markings to the bullet-proof, self-sealing, rail-mounted, rubber bladder petrol tanks.

The Bristol Hercules engine dressing and nacelle ('power egg') assembly area.

A visiting RAF aircrew being shown the fabric covering of Wellington tail-planes at Squires Gate by Charlie Boon, Chief Inspector, (ex-Weybridge).

Wellington production for Royal Air Force Coastal Command. Foreground: completed, equipped and fabric-covered fuselages. Centre: fin/rudder and other assemblies. Background right: fuselage/inner wing and power plant assembly, and left: complete aircraft final assembly. The Wellington was the most prolific British two-motor bomber of the Second World War and was in full-scale production at Chester throughout the conflict.

Left: Wellington Mk.XIV general reconnaissance version for RAF Coastal Command. Right: Wellington Mk.X trainer (the most prolific variant produced at Chester – 2,434) awaiting delivery. Wellington output from Chester peaked at around 130 aircraft a month (or around four and a half per day, seven days a week). The intensive production test flying at Hawarden was under the control of Maurice Hare who had transferred from Weybridge as Chief Production Test Pilot, and personally flew just over 3,000 aircraft on test at Chester, supported by only a few other ex-Weybridge test pilots and short-stay service pilots.

Wellington Production World Record

A dramatic highlight of the prodigious Chester-built Wellington output was a new World Record in 1944 for the fastest production assembly of an aircraft, the existing record being set by an American factory, which built a Douglas Boston twin-engined bomber in forty-eight hours. To boost national morale, the management and workers decided to attempt to assemble and fly a Wellington inside thirty hours of non-stop working. The mission was completed in twenty-four hours and forty-eight minutes from the start of assembly to lift-off, little over half the existing time. It was recorded by the Ministry of Information Crown Film Unit in an eighteen-minute documentary entitled *Workers Weekend – A Tribute to the Workers of the British Aircraft Industry*, narrated by an officer of the Royal Canadian Air Force to give it added authenticity in North America.

The normal works routine at Chester in wartime was 6am to 11pm, seven days a week, and about half of the 6,000 Chester workers were women. Between fifty and sixty workers were selected from the numerous volunteers to make the record attempt. Beginning at 9am on a Saturday morning, they also undertook to donate the bonus they would earn that day to the Red Cross Aid-to-Russia Fund. The rapid precision-fitting assembly of the main fuselage components and interior frames, systems and equipment resulted in the completed fuselage coming out of jig by 1.45pm for fabric covering and the application of nine coats of quick-drying, fabric-stretching and weather-proofing dope. Assembly of the complete airframe – with wings, petrol tanks, tail unit surfaces, bomb-beam and engine/nacelle 'power-eggs' that had been simultaneously pre-assembled and delivered from elsewhere in the factory – was complete by 6.15pm. The nightshift started at 8pm, the propellers, gun turrets and landing gear were fitted by 10.30pm, the RAF roundels painted on at 3am on Sunday morning, and the aircraft was towed from the assembly line to the engine-running shed twenty minutes later. By 6.15am, twenty-one hours and fifteen minutes into the record attempt, the aircraft was a complete fighting unit and saw the first dawn of its lifetime. After two hours of final adjustments and inspection, the fully-assembled, equipped, checked and tested aircraft was rolled out onto the airfield ready for take-off at 8.50am, just ten minutes short of a day. With ex-RAF test pilot, Gerald Whinney, at the controls, Chester-built Wellington Mk.X LN514, lifted off the Hawarden aerodrome after just twenty-four hours and forty-eight minutes of non-stop working from the start of the assembly and almost halved the world-record time for the fastest construction time for a bomber aircraft. At 7.45pm it was flown by a ferry pilot to its operational base.

Wartime Royal Visits

Above:
HM King George VI, appropriately dressed in Royal Air Force uniform, and Queen Elizabeth (the Queen Mother of today) visiting the Chester factory on 15 July 1942 escorted by Bernard Duncan, Superintendent. The Wellington Mk. III BJ708, which they saw take off, was delivered to No.38 Maintenance Unit (MU) RAF at Llandow the following day, before being issued to No.75 Squadron, but was lost on operations on 27 August 1942.

The first wartime Royal visit was by HRH The Duke of Kent, then an RAF Group Captain, who made a tour of the RAF Hawarden camp and No.7 Operational Training Unit (OTU) on 24 September 1940. He made a second visit on 7 May 1942.

Left:
HM King George VI taking a keen interest in the work of Olive North, one of the many skilled female workers at Chester, on the assembly of the Wellington front fuselage and nose gun turret ring.

The Broughton Wellington

The naming of 'The Broughton Wellington' Mk. IC R1333 – proudly bearing the Welsh Red Dragon emblem and inscription under the cockpit – on 7 November 1940. It was presented to the Royal Air Force through donations from the 'Workers and Co-operators' of Vickers-Chester, the Bristol Aeroplane Company (for the Pegasus engines) and their then 350 subcontractors. Together, they had donated £15,300 of the total cost of £20,000 with Vickers Ltd subscribing the balance of £4,700 (at a time when average shop floor weekly earnings were around £2.10s (£2.50).

In the letter to Lord Beaverbrook, the ebullient Minister of Aircraft Production, which accompanied the cheque, Duncan wrote, 'There is a natural desire that the gift should take the form of a particular Wellington allotted to an operational squadron soon after completion at the factory; also that it should be known as 'The Broughton Wellington' and the aircraft about to commence assembly at these works under the serial number R1333 should be the chosen one. By this means we should be able to keep the subscribers informed of some operations in which the aircraft took part and so maintain interest and enthusiasm.' However, despite this most commendable gesture, the aircraft did not actually see active service. After transfer to No.48 MU on the Hawarden aerodrome, and being damaged during the air raid directed at Coventry on 14 November 1940, a week after the hand-over, it was delivered to 99 Squadron at Newmarket, Cambridgeshire, on 1 December 1940. Unfortunately it crashed there on take-off on 18 December and was burnt out, killing the rear gunner. A replacement aircraft of the same name, R1516, was delivered to the No. (Czecho-Slovak) 311 Squadron and later lost on operations.

Left to right: Tommy Lucke, Vickers Test Pilot; Gordon Montgomery, Works Manager; Bernard Duncan, Superintendent; and Miss Scott, a senior secretary who had worked at Vickers House in London before the war, who cut the ribbon.

Other Notable Chester-Built Wellingtons

Wellington Mk. IC R1296, built at Chester in late 1940, which featured in the famous cinematograph film Target for Tonight, produced by the Crown Film Unit with the full co-operation of the Royal Air Force. Every member of the cast was played by actual members of the service carrying out their normal duties.

It was the Chester-built Wellington Mk. IC, L7818, on which Sergeant James Allen Ward, of No.75 (New Zealand) Squadron and second pilot of the aircraft, won the Victoria Cross on the night of 7 July 1941. Returning from an attack on Munster in Germany, the aircraft was attacked from beneath by a Messerschmitt 110 night fighter, setting fire to the starboard engine. Courageously, Ward crawled along the wing, kicking foot-holes in the fabric, and smothered the fire with an engine cover. By this heroic action, he enabled the aircraft to get home safely.

Chester factory workers must have been dismayed to learn that one of their aircraft, L7788, which crash-landed in Holland, was repaired and flown in German Luftwaffe markings!

Wellington T.Mk.X LN715, originally built at Chester and later adapted to be powered by the Rolls-Royce Dart propeller-turbine engines in association with the development of the Vickers Viscount, Britain's most successful post-war airliner, and first flown in this configuration on 10 June 1948.

To help maintain the required Wellington production rate at the Chester factory, in 1941 it became necessary to open a satellite assembly plant at Byley, twenty-five miles to the east, near Middlewich in mid-Cheshire, and to the south-west of the Cranage aerodrome (by which name the plant was always known). The large grass airfield there was originally built to accommodate RAF No.2 School of Air Navigation from October 1940 but was later adapted for fighter squadrons and used as a main line of defence for Merseyside. The first Cranage Wellington was completed in September of that year. This general view of the exterior of the Byley facility is fronted by a large number of fuselage transport dollies and aircraft jacks.

The Wellington flight shed, a field away from the Cranage aerodrome – to where the many hundreds of aircraft produced at the main works at Byley were towed along the interconnecting taxiway for preparation for flight and delivery.

Operating as a mass-production assembly plant only, with the manufacture of all components being subcontracted, the Chester factory and the Cranage satellite relied on a number of other local supporting dispersal sites to facilitate this process. This twin gable-ended hangar example at Aston Hall was replicated at Dobs Hill, Ewloe and Kinnerton. All had camouflage applied to simulate the appearance of large agricultural barns.

Pre-assembly of the fabric-covered geodetic tailplanes and flying control surfaces at the Anchor Motors garage in Chester and typifying the multitude of small local subcontractors supporting the Chester and Cranage Wellington assembly lines.

Foreground: The last batch of Vickers Wellington fuselage production. Background: the first Vickers-built Avro 683 Lancaster B.Mk.1 four-engined heavy bombers in final assembly at Chester.

Right foreground: Wellington final assembly line with (upper left) Avro Lancaster assembly line.

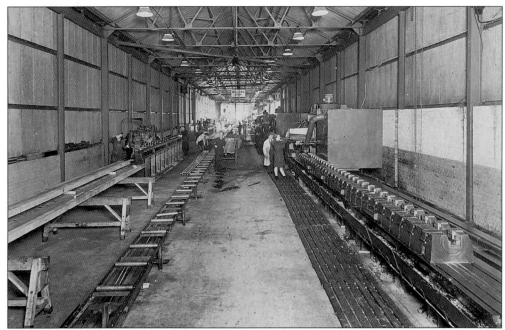

Wickman and Asquith machines in the Lancaster wing-spar milling shop. Vickers-Chester became responsible for construction of the centre wing/fuselage component for the Lancasters built there but with other major components being supplied by Avro's own main supply sources, notably Chadderton and Yeadon.

Lancaster centre wing-spar assembly.

52

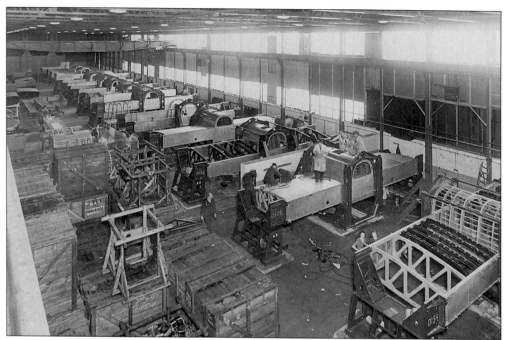

Lancaster centre wing and fuselage section assembly in the West Annexe. The conventional stressed-skin, monocoque-type fuselage structure and Warren-braced twin-spar wing construction of this aircraft was in total contrast to the geodetic-type structure of the Vickers Wellington.

Lancaster nose fuselage sections – incorporating the prone bomb-aiming, nose gun turret and cockpit windows. Roy Chadwick, the renowned Avro Chief Designer, had deliberately designed the Lancaster airframe to be broken down into large fully-equipped and road-transportable sections for ease and speed of dispersed production. (This method is common practice today, notably with international programmes such as the Airbus family, but now with fast long-distance aerial transport rather than the then shorter-distance road transport).

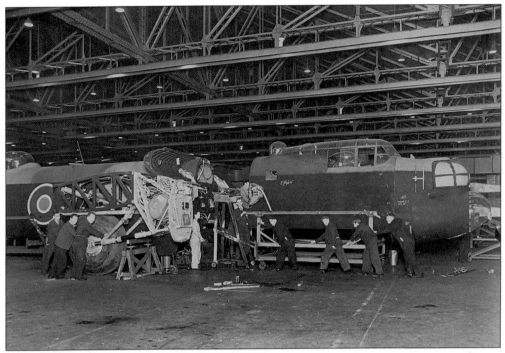

Mating of Lancaster nose and main fuselage sections.

Lancaster mainplane and control surface assembly.

Lancaster assembly line. Foreground: inner wing-to-centre wing/fuselage join-up. Centre: rear fuselage/centre wing assemblies. Background: Wellington fuselages undergoing fabric covering and internal equipping.

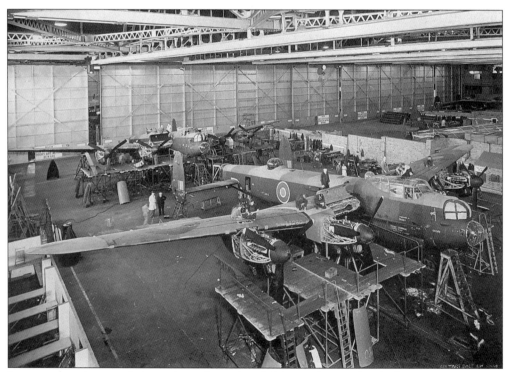

Lancasters nearing completion and transfer to the adjoining engine running shed.

Upper right: Lancaster assembly line paralleling continuing assembly of the Wellington. Vickers produced its first Lancaster in June 1944 and its 235th and last in August 1945, during which period a peak of thirty-six aircraft a month was reached.

The Avro Lincoln derivative of the Lancaster, originally known as the Lancaster IV and designed for operation against the Japanese. The abrupt ending of the Far East conflict in August 1945 meant that the Lincoln did not see action in the Second World War. Eleven of these Avro 'super-bombers' were assembled at Chester between June and August 1945 (together with the last twenty-three Chester-built Lancasters) from components supplied by Metropolitan Vickers at Trafford Park, Manchester. These were the last batch of the prodigious output of 5,786 aircraft to be built at Chester in exactly six years under the Vickers-Armstrongs management.

Three
The Blackpool Sibling

The second Vickers Wellington Shadow factory was located at Squires Gate airport on the Southern outskirts of Blackpool on the Fylde coast of Lancashire. Although operating completely autonomously, it had many similarities to that at Chester. Owned by the Government and operated by Vickers-Armstrongs, it was opened in 1940 and from October 1941 had a secondary assembly line at the Stanley Park municipal airport to the north-east of the town. However, unlike Chester, which was an assembly works only, Blackpool was laid out as a modern duplicate of the parent factory at Weybridge to manufacture the aircraft complete in all details. The entire aircraft output during the Vickers-Armstrongs management consisted of Wellingtons and totalled 3,406 aircraft, the first being completed in July 1940 and the last in October 1945. Eight were built between September the end of 1940, 199 during 1941, 615 in 1942, 930 in 1943, 1,125 in 1944, and 529 by October 1945, when production finally ceased. The Type breakdown was: fifty Mk. IC; 780 Mk. III; 1369 Mk.X; seventy-five Mk.XI; 802 Mk.XIII; 250Mk.XIV; and eighty Mk.XVIII. Again, as at Chester, 11,250 prefabricated houses were built by the Blackpool unit between September 1945 and April 1948, before Vickers finally vacated the plant in October 1949.

Wellington final assembly at the Vickers-Blackpool Shadow Factory built on the original historic Squires Gate aerodrome.

The massive machine shop at Blackpool, producer of the huge number of machined fittings used in the Wellington geodetic structural panel assemblies. It was estimated that each Wellington airframe incorporated 2,800 cross-over gussets and 'butterfly' fittings, 650 longeron gussets and 'wishbone' fittings, and 35,000 rivets. For the three-factory total of 11,460 aircraft built, and assuming the equivalent of a further 1,000 airframes for spares and repairs, means that the overall totals of the three types of fastenings were approximately 35million; 8.125million and 4.375billion respectively. These figures do not include special rivets/bolts in the main supporting structural framework and engine nacelles which could have increased the count by at least another 10 per cent.

The Wellington inner and outer wing fabric covering, stitching and doping area at the Blackpool works, populated exclusively by female workers as they were presumed especially adept at this tailoring-like work.

Highlights Of The Blackpool Shadow Factory

Alex Dunbar, General Manager of the Vickers Aircraft Section, decided on the location of the Vickers Wellington Shadow Factory at Blackpool after sites at Exeter and Doncaster had been studied. The Air Ministry contract for the construction and equipping of the new unit by the Ministry of Aircraft Production was issued on 30 December 1939, and the first sod cut on 3 January 1940. The Squires Gate aerodrome, on which the factory was built, was the scene of a famous flying meeting in October 1909 organised by Lord Northcliffe, proprietor of the Daily Mail (giving the present airport a longer history than any other in Britain). With good approaches and flat surroundings, at the outbreak of the Second World War it was seen as a good base for RAF Coastal Command. As at Chester, Wellington production began in a Bellman hangar pending completion of the Shadow Factory on the edge of the airfield. Started from Weybridge and not from Chester, as might be supposed, the first contract was for 500 Wellingtons, the first aircraft was rolled out in July 1940 and the first three production machines were delivered by the end of that month.

The factory was much bigger than Chester, and had its own machine shops and subcontractors. Production dispersal depots were opened in the area, including the Harrowside Bridge and Belle Vue garages, the Talbot and Devonshire Road bus depots, the Whitegate Drive tram depot and St. John's Market Hall. A secondary assembly line at Stanley Park was established on 26 October 1941.

A shortage of skilled labour was partly overcome by the transfer of personnel from the Vickers shipyard at Barrow-in-Furness. However, a serious hold-up in the production programme occurred on 9 August 1940. After the completion of the construction of the main works, the centre-section of the main aircraft erection bay collapsed, killing six men outright and injuring thirteen others, one fatally. The resulting disruption had a serious effect on the output of aircraft. The bombing and serious disruption of the Vickers-Weybridge factory a month later meant the MAP also instructed Vickers to disperse the test flying of all its experimental aircraft types to Blackpool. When full production resumed, a rate of fifteen to twenty aircraft a week was reached. At one time a contract was received for the production of 300 Vickers Warwick aircraft, but shortly afterwards cancelled. The manufacture of 11,250 AIROH prefabricated houses, the same number as at Chester, superseded that of aircraft. The first was completed in September 1945 and the contract completed in April 1948 – when the factory was finally vacated by Vickers on 18 October 1949 before resuming aircraft production for a period during the early 1950s by Hawkers with the Hunter jet fighter.

The administrative arrangements at Blackpool were generally similar to those at Chester. Seen here with the one-thousandth Blackpool-built Wellington ready for delivery in 1942 are, left to right: Frank White (Subcontract Manager), Les Webb (Production Manager), Sam Bower (Superintendent throughout the period of production by Vickers), Ernie Comley (Personnel Manager), and Teddy Major (Chief Inspector).

Vickers Wellington geodetic fuselage structures in progressive stages of completion at the Blackpool Squires Gate Shadow Factory.

Wellington wing production and fabric covering exhibiting the same kind of geodetic construction as the fuselage and other fixed flying control surfaces.

Wellington component sub-assembly manufacturing area showing, bottom to top: inner wings, control surfaces and outer wings, fuselages, and, top right: fuselage equipping and skinning.

Prototype vickers Wellington MK.IV R1220 built at Chester but seen here outside one of the two camouflaged shallow 'V' clerestory roofed hangers at squires gate (as shown on page 63) to where this aircraft had been flown for experimantal flight-testing.

A Blackpool-built, Bristol Hercules-powered, Vickers Wellington T.MK.XVIII NC869 fitted with a de Havilland Mosquito nose and modified to house a radar scanner for training night-fighter crews.

The first prototype pressurised-cabin, high-altitude Wellington Mk.V R3298, which was flown from the Vickers-Weybridge factory to Blackpool on 25 September 1940. After the disastrous bombing of the vulnerable Brooklands factory and airfield on 4 September 1940, the Government insisted that all test flying of Vickers experimental aircraft types be transferred to Squires Gate. However, it was not until late October 1940 that high-altitude flight up to 30,000ft and beyond was possible due to icing of the pilot's canopy dome.

The last Blackpool-built Wellington T.10 Trainer, RP590, and the very last of the 11,460 production Wellingtons built at Weybridge, Chester and Blackpool, making a final salute to the Squires Gate factory prior to being handed over to the Royal Air Force on 25 October 1945.
RAF Museum/Charles E. Brown Collection (6083-7)

The Blackpool Wellington Legacy

Most notable among the 3,406 Wellingtons built at the Blackpool Shadow Factory was Wellington Mk.X MF628 which, after a distinguished active display career, is today the only surviving complete Vickers Wellington and is displayed in the Bomber Command Collection at the Royal Air Force Museum at Hendon, North London. First flown at Blackpool on 9 May 1944, MF628 was initially allotted to No.18 Maintenance Unit (MU) RAF, at Dumfries, Scotland, on 11 May 1944 before being modified to T.Mk X standard by Boulton and Paul in 1948. Damaged during service as a navigational trainer, after a ten-month repair by Brooklands Aviation at Sywell, Northamptonshire, between December 1951 and October 1952, it was tranferred to 19 MU RAF at St. Athan, South Wales, before taking part in the 'Fifty Years of Flying' display at the Royal Aeronautical Society's (RAeS) Garden Party at de Havilland Hatfield on 14 June 1953.

After flying over Lake Windermere in August 1954 to appear in the film, *The Dambusters*, MF628 was sold to Vickers-Armstrongs at Weybridge on 24 January 1955 and flown from St. Athan to Wisley in the last ever Wellington flight. Vickers then presented the aircraft to the RAeS at the Society's Garden Party at Wisley on 15 July 1956. After several more moves during the next six years, this historic aircraft was handed over by the RAeS on permanent loan to the Ministry of Defence in mid-1964. Refurbished to display standard at St Athan, and exhibited at RAF Abingdon before HM The Queen on 14 June 1968 to commemorate the fiftieth anniversary of the RAF, MF628 was ultimately moved on 26 October 1971 to its current proud resting place at the RAF Museum at Hendon. The only other surviving Wellington, and the only one to have seen operational service during the Second World War is the Weybridge-built aircraft that was dramatically recovered from Loch Ness in Scotland in September 1985. It has since been splendidly restored and preserved at the Brooklands Museum with the partly covered airframe readily exhibiting the distinctive geodetic type construction.

Four
From Bomber to Bungalow
A Peace Dividend

During the Second World War, the manpower in the house-building sector diminished dramatically as the personnel were required in the armed forces and the war industries, and later to repair bomb-damaged homes. By the end of the war, timber and bricks were in short supply, meaning a serious delay in the building of much needed conventional homes. This prompted the preparation of an AIROH (Aircraft Industry Research on Housing) pre-fabricated aluminium and concrete house design for large-scale production as a temporary solution to the housing deficit.

The Vickers shadow factories at Chester and Blackpool were each contracted by the Government to build 11,250 of these ingenious constructions. The conversion of aircraft factories for the construction of these 'prefabs' was a major undertaking but expeditiously fulfilled. Moreover, the utility of this radically different output provided a vital continuity in employment as well as a valuable contribution to the regeneration of peacetime Britain. More than 150,000 of these homes were built and erected nation-wide, with funding and materials from the US as part of the post-war Marshall Aid Plan. Valued at £1,200 each in 1947, they were typically rented at fifteen shillings per week, and those remaining might well now sell for £35,000.

A typical AIROH utilitarian-design, cottage-style pre-fabricated home, fully-assembled, equipped and occupied. Each unit was built in modular form and comprised a living room, two bedrooms, fitted kitchen (not previously seen in the UK, with worktops in pressed steel and cream and green enamel), bathroom, toilet, and hall. Each had a front entrance, the internal doors connected directly with the hall, with the exception of the kitchen which was entered either from the living room or from outside.

A corner of the foam cement plant with the cement silo fronted by the receiving building.

The partition wall casting platform in the foam cement plant. The outer walls were sandwiched between two sheets of light alloy and baked cement.

Internal partition wall plasterboard application area and upending gear in the foam cement plant.

Foreground: the partition wall undercoat paint plant with the dripping and draining tanks. The production line moved at $1\frac{1}{2}$ft per minute. (This process plant was an interesting portent of the huge completely-automated Airbus wing skin panel treatment plant that was installed at Broughton during 1999).

The kitchen and bathroom unit (KBU) stoving oven, typifying the large-scale specialised machinery that had to be installed for the production of pre-fabricated houses instead of aircraft

Floor assembly in the south and main shop floor areas of the Chester factory, indicating the value of adaptation of the extensive overhead crane rails originally installed for the movement of large aircraft components.

Wall assembly with an immense amount of materials being handled.

Assembly of the Warren-braced hip roof support modules.

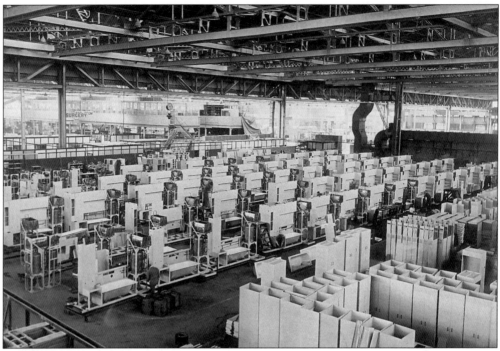

Assembly of the modular KBU and cupboard unit structures.

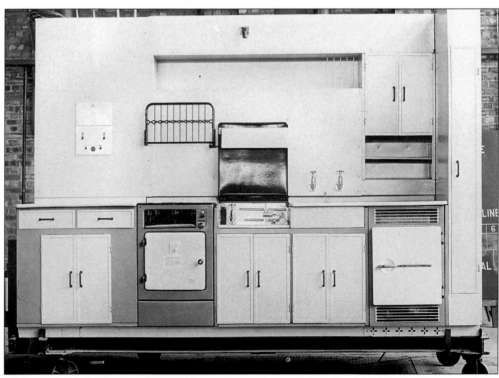

Kitchen side of the finished KBU with a modern, space-efficient fitted kitchen, with oven and gas refrigerator. This unit was an integral back-to-back module separating the two rooms of the dwelling.

Bathroom side of the KBU with; right: built-in lagged main and, hot water tanks.

KBU module assembly lines, with floor, and, foreground: wall and external window frame installations moving on wheeled platforms and floor track runners for automobile-style line production.

Attachment of the roof structures using a 'Kings' overhead hoist.

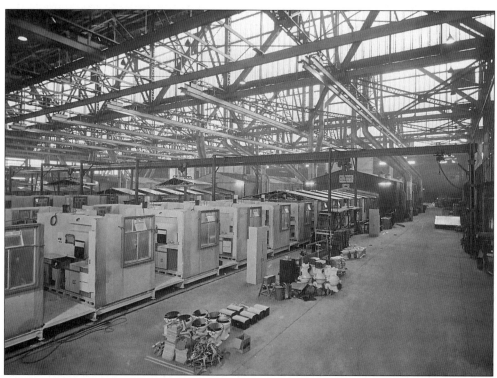

Main assembly lines (kitchen side of the KBU), moving towards attachment of the roof modules and into the final external treatment process chambers.

Assembled house modules moving into the external treatment process chambers.

One of the four fully-equipped ex-works house modules being loaded for transport by contract carrier to the erection site. Each house left the factory in four sections, completely equipped and painted, on their own trailers. Seven men could then erect it in two hours on a pre-prepared foundation. A few hours were then required for the connection of public utilities.

Despatch area with house sections loaded with temporary hessian transport covers at the open ends.

Background: prefabricated house section leaving the Chester factory site past the main canteen and works transport building. By the end of 1945, a few of these inventive and fully-functional domestic dwellings had been built but production quickly gathered momentum and a maximum weekly output of 170 complete units was soon reached. Originally intended to last only ten years, hundreds of them remain in occupation today, more than fifty years after they were originally built – testament to both the original concept and the quality of the ex-aircraft workmanship. In 1998 several surviving prefab housing sites were given Grade II listing for posterity by English Heritage, most notably a row of sixteen original prefabricated bungalows at Wake Green Road in the Hall Green suburb of Birmingham, installed in 1946-1947, still occupied and little changed.

Five
The de Havilland Dynasty

The availability of the Chester facility in 1948 enabled the de Havilland Aircraft Company (DH) to fulfil the burgeoning export order-book for its Vampire jet fighter and Dove light transport aircraft. It became the the main outlet for all designs in the prolific DH dynasty during the next twelve years of the company's operation of the factory. For five years from 1960 the DH name continued on the masthead under the Hawker Siddeley regime as the HS de Havilland Division. The open-plan floor enabled large-scale production lines of numerous types to be handled simultaneously. The transfer of the completed units to the pre-flight area was made via the inherited factory-wide overhead craneage system. However the smaller Chipmunk and Beaver and the larger, more complex and slower-moving Comet jetliner were built on conventional nose-to-tail and herring- bone assembly lines. 3,548 aircraft of eight DH types were produced between mid-1948 and the end of 1959 – Mosquito (81), Hornet (149), Vampire and Vampire Trainer (1,236), Chipmunk (889), Dove (209), Venom (834), Heron (129) and Comet (13).

The new DH name above the entrance to the Chester site management and administration block in 1948, with the twin stone-carved interlocked 'V-A' (Vickers-Armstrongs) logos of its predecessor still in situ.

Assembly at Chester of the first laminated wooden ('pea-pod') fuselage half-shells for the famous D.H.98 Mosquito fighter-bomber (which was dubbed 'The Wooden Wonder') in mid-May 1948. The Mosquito was the first DH manufacturing programme to be undertaken there and was the end phase of the overall Mosquito production run. The Chester output consisted of sixty-seven N.F.Mk.38 Night Fighters and fourteen T.R.Mk.37 Sea Mosquitos completed between September 1948 and November 1950. Most of the former were for Yugoslavia. The latter version served with 703 Squadron at Solent area bases and provided the Naval Flight of the RAF's Air-Sea Warfare Development Unit.

VT670, the first Chester-built Mosquito N.F.Mk.38 taxiing past the last of the Vickers Wellingtons for its first flight from the Hawarden aerodrome on 30 September 1948. The last Chester-built Mosquito was VX916, an N.F.Mk.38, which was first flown on 15 November 1950 and was the 7,781st and last ever example of this extraordinarily successful aircraft.

The D.H.103 Hornet production line in 1950, of which variety 149 were built at Chester between 1949 and 1952. The sleek single-seat Hornet was the World's fastest piston-engined aircraft and was intended for use in the Japanese campaign but did not see war service. While the fuselage was of similar construction to the Mosquito, the wing was of composite construction with a double upper skin of plywood and an under-skin of light alloy, with composite wooden spars, assembled in what became known as the 'Redux' bonding process – which featured widely in subsequent all-metal DH designs.

A major spares organisation for the whole of the DH civil and military aircraft family was set up at Chester in 1950, with a world-wide distribution network. A service department was also established there to handle factory repair of military aircraft and provide after-sales service to some of the civil aircraft customers. This was later complemented by a nucleus design staff to deal with the customisation of civil aircraft.

The first Chester-built D.H.100 Vampire twin-boom jet fighter, an FB.5 (with DH production serial V-0080, later registered VZ841) being rolled out for flight testing prior to delivery to the RAF on 30 March 1951 Top left: Mosquitos and top right: Hornets, in front of the Chester flight shed awaiting delivery. This variant became the most common in RAF service, changing the Vampire's role from one of intercepter, to support ground-attack fighter-bomber. Many squadrons were stationed with the Second Tactical Air Force (TAF) in Germany.

Front row: pre-delivery line-up in early 1951 of the last three Hornets built at Chester; middle row: D.H.C.1 Chipmunk fully-aerobatic pilot trainers for the RAF; bottom left: Vampires of an evaluation batch for Norway. Chester-built Chipmunks were exported to over forty overseas air arms.

Production of Vampire cockpit/nose stressed-skin fuselage sections with laminated cedar plywood-balsa-plywood carapace shell structures cemented under pressure, in the form of the fuselages of the pre-war D.H.91 Albatross airliner and the wartime Mosquito. Many Vampire fuselages were built by Fairey Aviation at Heaton Chapel, Stockport, Cheshire, and transported by road to Chester for completion.

Assembly of the Vampire cockpit/nose undercarriage and gun bay to the centre fuselage/wing section.

Large-scale production of the D.H.115 two-seat Vampire Trainer. Upper left: the D.H.C.1 Chipmunk assembly line. Chester contributed the largest output of the 4,366 Vampires produced by twelve manufacturers in the UK, Australia, France, India, Italy and Switzerland. Of the 3,269 Vampires built in the UK, Chester built a total of 1,244 of all types between 1949 and 1963: J28.B for Sweden (297); FB.5 (87); FB.9 (705); NF.10 (55); FB.52 (114); FB.54 (12) and 424 T.11 and T.55 Trainers. The Vampire was the first British aircraft to exceed 500mph.

Chester-built Vampire T.11 Trainers WZ456, WZ457 and WZ453, some of the most widely used advanced jet fighter trainers. The T.11 became the standard jet trainer of the RAF and the Royal Navy. The T.55 version was supplied from Chester to more than twenty air forces: Austria, Burma, Ceylon, Chile, Egypt, Eire, Finland, India, Indonesia, Iraq, Italy, Japan, Lebanon, New Zealand, Norway, South Africa, Sweden, Switzerland, Syria and Venezuela. 117 Vampires were refurbished there. The RAF continued with the type until 1970 and the Swiss Air Force until 1992.

Vampire Trainer being transported from the Chester main assembly hall to the paint shop and flight preparation area on a 7½ton 'Royce' overhead crane.

D.H.112 Venom N.F.Mk.2 night fighters – using the basic Venom single-seat airframe with a two-seat side-by-side airframe and AI (airborne interception) radar in an extended nose – on the Hawarden aerodrome awaiting delivery to No.23 Squadron Royal Air Force at Coltishall, Norfolk. The Venom fighter-bomber, night fighter and Sea Venom constituted the second generation of the distinctive D.H. twin-boom formula. The more powerful D.H. Ghost engine replaced the D.H. Goblin and Rolls-Royce Nene of the Vampire – chosen to shorten the jet tailpipe and reduce power losses from the relatively primitive jet engines then under development. Chester also built sixty-two Venom NF.51s, known as J.33s, with Fairey at Ringway, Manchester, between 1952 and 1957 for the Royal Swedish Air Force. They had Ghost engines, produced under licence in Sweden by Svenska Flygmotor as the RM2A and shipped to Chester for installation.

Final assembly of some of the ninety-nine Sea Venom F.A.W. Mk.21s built at Chester in 1955 and 1956. WL813, one of the seventy-three Venom F.B.Mk.4s, made the first air firings of the DH Firestreak guided missile at Aberporth, destroying a Fairey Firefly drone in the process. The Mk.21, XG607, then conducted the service trials of the missile with No.700 Naval Trials Squadron at Ford, Hampshire. Three Chester-built Mk.21s were used for the Royal Navy firing trials which began in December 1958 when three Sea Venom F.A.W Mk.21s were catapulted from HMS Victorious in the Mediterranean in preparation for the introduction of the Firestreak on the succeeding Sea Vixen aircraft, recording 80 per cent direct hits on Malta-based Fireflies.

WD327, the 147th Chester-built D.H.C.1 Chipmunk pilot trainer for the RAF returning from a test flight. It was designed as a successor to the venerable DH71 Tiger Moth, and adopted for the Royal Air Force Volunteer Reserve (RAFVR) flying schools to Specification 8/48 as the T.Mk10. The export demand, with the RAF order for 740 aircraft, was large enough to justify production in the UK for those countries which found it easier to pay in Sterling. UK production of 1,000 Chipmunks was begun in 1949 at Hatfield where 111 were built; the rest were built at Chester, the 1,000th (WZ864) coming off the line on 25 February 1956. A number were also finished by a subcontractor, Hooton Aero Engineering, at nearby Hooton Park.

Centre: large-scale refurbishment of RAF Vampire Trainers in 1955 – hence the apparently random serials.
Right: the Royal Navy Sea Venom production track. Chester refurbished 177 Vampires between 1954 and
1964, and handled eighty-seven Venom overhauls/resales between 1964 and 1968. 1,241 aircraft were
overhauled, refurbished and repaired between 1952 and 1971.

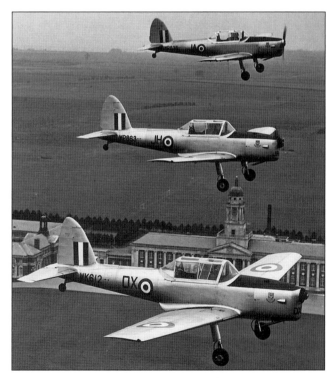

Chester-built Chipmunks
WK612, WP863, and WP836
over their principal Alma Mater,
the renowned Royal Air Force
College at Cranwell,
Lincolnshire.

Royal Choice: HRH Prince Charles learning to fly in August 1968 at RAF Tangmere, Sussex, in the Chester-built Day-Glo red Chipmunk T.10 WP903 of The Queen's Flight, under the instruction of Sqn Ldr Philip Pinney of New Zealand serving with the RAF Central Flying School at Little Rissington. Charles received his 'wings' in March 1969. HRH Prince Philip had learned to fly at White Waltham with Chester-built Chipmunks WP861 and WP912, going solo in the latter, now preserved at the RAF Museum Cosford, Shropshire, on 20 December 1952, and using WP912 shortly thereafter and WF848 in 1954/55. Quickly becoming a proficient pilot on the Chipmunk, Prince Philip progressed to the Chester-built twin- and four-engined D.H. Dove and Heron and other aircraft. WP903 was used by Prince Philip and then for instructional flying by the Duke of Kent, Prince Michael of Kent and the late Prince William of Gloucester. In early 1979, WP904 was used to teach Prince Andrew for his Royal Naval Pilot Graduation.

A Chester-built Dove in military guise as the Royal Air Force Devon C.Mk.I VP978 which served successively with the British Air Attaché in Teheran, Communications Flight in Iraq, Air Attaché in Saigon, conversion to a C.Mk.2, with Nos.21 and 26 Squadrons (temporarily carrying the civil registration G-ALYO when serving on Air Attache duties) and typifying the versatility of the adaptation of the civil Dove to serve in foreign capitals with both the RAF and the Royal Navy (as the Devon and Sea Devon respectively).

Final assembly of the D.H.104 Dove light twin-engined transport in 1955 – one of six different DH aircraft types in high-rate production at Chester at that time. So popular was the Dove that production continued there for seventeen years, from 1951 to 1967, and to a total of 244 units. They were all powered by successive versions of the D.H. Gypsy Queen engine, which saw service throughout the world – in civil operation with airlines, business companies and individual owners; and in government and military service with airforce and naval VIP and staff transport arms.

Personal Choice: The specially adapted Chester-built Chipmunk G-APOY (ex-RAF WZ867), owned by the Airways Aero Association and piloted by (later Sir) Peter G. Masefield. It was one of four such aircraft modified in 1959 for civil operation by Bristol Aircraft at Filton (Bristol) at his instigation whilst managing director of the company . It incorporated a single-piece canopy (as used for the DH Canada-built Chipmunk T.Mk20 for the Royal Canadian Air Force), navigation beacon and wheel spats. He used a similar aircraft (appropriately registered G-APGM) to commute between Redhill aerodrome, near his Surrey home at Reigate, and the Filton factory airfield. Many ex-Service Chipmunks still fly in private hands.

Dove production for the Royal New Zealand Navy and North American customers. The Chester Dove production output made a valuable contribution to the national export drive during the 1950s and 1960s.

Dove Series 8 G-ARBE, one of the last ten of these aircraft built at Chester, in the house livery of the parent company, Hawker Siddeley Aviation, from 1965, at London-Gatwick airport in June 1970. Dove 8s G-ASMG and G-AREA also served successively with Hawker Siddeley Aviation and British Aerospace. This latter aircraft is now preserved at the de Havilland Heritage Centre at London Colney, Hertfordshire.

Series production at Chester of the enlarged four-engined derivative of the Dove – the D.H.114 Heron fourteen to seventeen seat feeder-liner – forming a modern successor to the pre-war D.H. 86 four-engined biplane transport. After the first seven Herons had been built at Hatfield, the last being the prototype Heron 2 with a retractable undercarriage, the production line was transferred to Chester, where 140 Herons were built for VIP, commercial and military service between 1953 and 1967.

Chester-built Heron 2 G-ANCJ powered by DH Gypsy Queen Series 30 Mk.2 un-supercharged engines and used as a de Havilland sales demonstration and company transport aircraft. The first Chester-built Heron Series 2 was handed over to the French carrier Union Aeromaritime de Transportes (UAT) on 7 March 1953 and was later sold in the Congo and used by the ill-fated President Tshombe of Katanga.

Royal Choice: XM296, one of the fleet of four VIP Herons chosen by HM The Queen's Flight – built and serviced for many years at Chester and based at Royal Air Force Benson, Oxfordshire – at the opening of London-Gatwick Airport by the Queen on 9 June 1958. The other three aircraft were: XM295, XM375 and XM391. These aircraft were flown extensively by HRH Prince Philip. They also made numerous overseas tours between 1958 and 1963 with the British Royal Family, numerous overseas monarchs and presidents, and home and overseas political leaders. They flew a total distance of two million miles and a total flying time of 13,400 hours. Several overseas potentates also chose the Heron for their personal use.

Airline Choice: Chester-built Heron 2 G-ANLN serving with Jersey Airlines as Duchess of Guernsey, illustrating the commercial popularity of the Heron. This aircraft was one of two in the fleet (together with three ex-West African Airways aircraft) later converted (at D.H. Leavesden) for military communications under the designation Sea Heron C.Mk.20 XR441 to serve with 781 Squadron Royal Navy at Lee-on-Solent. In May 1961 the Royal Navy decided to replace five Sea Devons (the naval version of the Dove) with a similar number of ex-civil Heron 2s.

Business Choice: Chester-built Heron 2 G-AOGW Golf Whiskey used for many years as part of the company transport aircraft fleet of Vickers-Armstrongs (Aircraft) at Weybridge, Surrey, and based at the nearby flight test airfield at Wisley. The Heron proved to be an ideal choice as a corporate communications aircraft and was a logical progression from the Dove. Chester-built Herons were notably used by Rolls-Royce, English Electric, Ferranti, British Aircraft Corporation (BAC) and the Royal Aircraft Establishment (RAE) in the UK; and overseas by Phillips of the Netherlands and Fiat of Italy.

Military Choice: Chester-built Royal Navy Sea Heron C.Mk.4 XR391 serving with 781 Squadron. Military Herons, with their 'four-engine safety' attribute, were also supplied from Chester to the RAF and the communications flights of the air forces of Saudi Arabia, South Africa, Ceylon, Iraq, West Germany, Jordan and Ghana. These included one once maintained as the personal aircraft of King Feisal of Iraq, and those frequently used by King Hussein of Jordan and President Nkrumah Of Ghana.

The first Chester-built D.H.106 Comet 2 fuselage, with the original rectangular-shaped passenger windows, nearing completion in midst of large-scale production of Vampire Trainer cockpit/nose fuselage sections (with a Chipmunk, lower left).

A Comet 2 fuselage, together with other airframe components and tooling, being shipped by barge to Chester via Liverpool docks. It was en route from the moribund second Comet assembly line started by Short Brothers at Belfast but abandoned when all Comet production was halted after the early in-service Comet accidents. A number of other Royal Air Force production contracts were directed to Northern Ireland, e.g., the Bristol Britannia, Vickers-Supermarine Swift (not activated) and a front fuselage section of the Vickers VC10.

The Belfast-built Comet 2 fuselage being towed past the Tudor-faced architecture of the historic Roman-walled city of Chester and thus honouring the name by which the Broughton aircraft factory and Hawarden aerodrome complex became known over much of their history.

Despite the suspension of the initial Chester Comet 2 line in 1954, reworked Comet 1A and 2 airframes for the RAF and others were completed at Chester between 1955 and 1957 and served as an educational operation for the building of the strengthened and enlarged Comet 4. Scaffolding was used for access to these much bigger aircraft untill the construction of two-deck staging. Production of the diminutive Vampire, Venom, Dove and Heron was to continue throughout the next decade alongside the radically different manufacturing scale and techniques of the Comet.

The second of two Comet 1A VIP transports, VC5302, built at Hatfield for the Royal Canadian Air Force Air Transport Command in 1953, being re-accepted outside the Chester flight shed on 26 September 1957 after major fuselage strengthening and conversion to Comet IXB standard. Used by the Queen and Prince Philip during their visit to Canada in 1959, this aircraft remained in service until October 1964. The last of the original Comet 1s, these two aircraft were used to simulate high-speed jet bomber penetration to test the Canadian defences. The first built was also the first jet transport to cross the Atlantic when delivered to No.412 squadron at Rockcliffe, Ottawa on 29 May 1953. Two Hatfield-built Comet 1As, acquired by the UK Ministry of Supply in 1954, were similarly modified at Chester in 1958 for experimental use by the Royal Aircraft Establishment (RAE).

The Royal Air Force Comet 2R XK663, originally intended for commercial operation with British Overseas Airways Corporation (BOAC). It was one of fifteen Hatfield-built Comet 2s later converted at Chester to 2.R and T.2 standards for the RAF, for transport and training roles, and delivered to No.192 and No.216 Squadrons respectively between 1956 and 1958. These aircraft were complemented by the only Chester-built RAF Comet 2 XK716, named 'Cepheus', and delivered to No.216 Squadron in May 1957 – which became the world's first military jet transport squadron. The RAF Comet 2 was also the world's first 'stretched' jet airliner. The XK716 was the only Comet built from assemblies produced by Shorts of Belfast, and also the last Comet 2 before both Hatfield and Chester began production of the enlarged Comet 4. Twenty-five Comet 2s were to have been built at Chester but the BOAC in-service disasters led to them all being cancelled, although four had been partially completed.

The first Chester-built Comet 4 fuselages – notably incorporating oval passenger window cut-outs instead of the rectangular-shaped ones of the original Comet design – nearing the proof-pressure testing stage in early 1958. The Chester factory also became responsible for the manufacture of a number of large sub-assemblies for the Hatfield Comet 4 production line.

The original factory-wide overhead crane system at the Chester factory adapted to lift and greatly simplify the mating of the complete Comet 4 fuselage and wing structures (here for the second East African Airways aircraft). This demonstrates the well-matched scale of the Chester assembly area to that of the largest aircraft type to be built at Chester. The crane drivers (who were invariably female) quickly became adept at swinging these very large loads with ease for precise positioning.

The G-APDE, the first Chester-built Comet 4 for BOAC, and the fourth of the airline's order of twenty, being made ready for its first flight. It was taken from the Hawarden aerodrome, on 20 September 1958, to Hatfield for flight testing, customer acceptance and delivery on 2 October – two days before two Hatfield-built BOAC Comet 4s inaugurated the historic world's first transatlantic jet airliner service between London and New York. G-APDE notably operated the first jet service between London and Singapore.

G-APDR, the seventh Chester-built Comet 4 for BOAC leaving the flight shed for its first flight on 9 July 1959 and delivery eleven days later on 20 July 1959. Five more BOAC aircraft were delivered by the end of that year. Altogether, twelve Comets 4s were built at Chester for BOAC, the last being G-APDJ delivered on 11 January 1960, two months ahead of schedule. Together with the single Chester-built Comet C2 for the Royal Air Force, this brought the total Chester Comet output to thirteen before the de Havilland Aircraft Co. became part of the Hawker Siddeley Group in January 1960. However, all subsequent Comets built at both Hatfield and Chester continued to carry the D.H. prefix with the final delivery on 26 February 1964.

Six

Under the Hawker Siddeley Banner

Although the Government-directed consolidation of the UK aircraft industry in January 1960 resulted in the pioneering de Havilland company being subsumed into the Hawker Siddeley Group (HSG), the DH name was to prevail for a further four-and-a-half years as the HS de Havilland Division until the firm became fully absorbed into Hawker Siddeley Aviation (HSA) in June 1965.

During the 17 years of operation under the Hawker Siddeley banner, the Chester factory saw the final stages of the production of the DH Vampire Trainer (8), Dove (35) and Heron (11); and the production of the Comet 4 (28), Sea Vixen (30) and DH Canada Beaver (46). However, the three great sustainers at Chester were the successful DH/HS 125 business jet, the main airframe components for the Comet-based HS801 Nimrod maritime patrol aircraft (41), and the private contract suppy of wing structures for the European Airbus airliner consortium. Completing the first wing set in November 1971, the Government-directed formation of British Aerospace (BAe) in April 1977 meant that HSA bequeathed the new company this increasingly successful aerostructures programme and continuing high-volume output of the 125. and, most importantly, the exceptional capability and record of the Chester site.

The facade of the original management and administration block fronting the main Chester factory and up-dated for the third time in mid-1963 to incorporate the Hawker Siddeley Aviation Limited title. Its previous incarnations were: The Hawker Siddeley Group, in place from 1960 to 1963, de Havilland Aircraft Company from mid-1948, and the founding Vickers-Armstrongs Limited from 1939.

Foreground: forward section of the Comet 4 fuselage panel manufacture (situated immediately behind the flight deck, assembled separately). The third section is that for the over-wing fuselage component and incorporates the large over-wing emergency escape hatches, as well as the redesigned oval passenger window apertures.

Civil Comet 4 final assembly in full swing on two assembly lines in the early 1960s with the first of a fleet of nine Comet 4Cs for Misrair/United Arab Airlines, the national flag carrier of Egypt, at the livery application stage. Built between May 1960 and February 1964, this airline's Comet fleet was flown extensively throughout the Middle East, the Mediterranean, Europe and North Africa. Other civil Comet 4s built at Chester under the Hawker Siddeley banner between July 1960 and February 1964 included: three each for East African Airways and Aerolineas Argentinas; three (longer-bodied) 4Cs for Middle East Airlines; two short-haul (also longer-bodied) 4Bs for British European Airways; and two 4Cs for Kuwait Airways.

Military Comet C.4 production for Royal Air Force Transport Command, built to serve with No.216 Squadron. In 1975 all were acquired by the UK independent airline, Dan-Air of London, to become the last of a total fleet of forty-eight pre-owned Comet 4s of all types (including some for spares) which enabled the airline to become the largest Comet operator.

Royal Air Force Comet C.4XR395, the first of the five such aircraft – which in civil guise became G-BDIT. XR398. The fourth became G-BDIW and made the final Comet commercial flight on 9 November 1980, before retirement in February 1981 for preservation at Dusseldorf, Germany. Other ex-Dan-Air Comets have notably survived at Duxford, Cambridgeshire (G-APDA); Wroughton, Wiltshire (G-APYD); and East Fortune, Lothian, Scotland (G-BDIX). Dan-Air Comet 4s carried more than eight million passengers between 1966 and 1980.

Chester-built Comet 4C XS235 which was the last Comet flying, up to 14 March 1997, before being acquired by de Havilland Heritage. Here it is saluting its original home at Broughton while being ferried to Bruntingthorpe, Leicestershire, on 30 October 1997, to be maintained by the British Aviation Heritage Group, with the hope that it will eventually be made airworthy again. On its first flight to Hatfield on 26 September 1963 to be fitted out as a flying radio and avionics laboratory, it was named Canopus. During its thirty years in service, it served with the Blind Landing Experimental Unit (BLEU), Bedford; the Aircraft and Armament Experimental Establishment (A&AEE) at Boscombe Down; and the Defence Test Evaluation Organisation (DETO) at Farnborough.

Centre: John Cunningham, the famous D.H. Chief Test Pilot with, left: his deputy, Pat Fillingham and, right: Tony Fairbrother, Chief Flight Test Engineer, with the penultimate production Comet at Chester. Originally a Comet 4C, built 'on spec' without a committed customer, it was first flown (registered G-5-1) on 25 October 1965. After storage, this aircraft was reactivated with the military serial XV147 and flown to HSA Woodford (Manchester) to become the aerodynamic development prototype for the HS801 Nimrod maritime patrol aircraft (but retaining its Rolls-Royce Avon engines) and first flown there in this guise on 31 July 1967. (In 1953, Fillingham had flown a D.H. Chipmunk to victory in the King's Cup Air Race averaging 140mph).

De Havilland Canada D.H.C.2 Beaver AL Mk.1 XP770 high-wing STOL, 'half-ton-truck' bush transport aircraft, one of the original order for thirty-six of these aircraft assembled at HSA Chester in 1961 and 1962 for the British Army Air Corps, plus four for the Ministry of Defence, awaiting delivery on the ramp outside the flight shed. Although adequately equipped for reconnaissance, the British Army Air Corps required the Beaver for liaison duties and casualty evacuation in conditions of field support where the availability of prepared surfaces could not be counted on. A supplementary order for a further six Army aircraft was completed in 1962. The four aircraft ex-MoD with RAF serials XR213-216 were allotted to the Muscat and Oman Air Force.

Air-testing of D.H.C.2 Beaver XP769, one of the forty-six Chester-assembled Beaver light military transport aircraft built from kits of the main components manufactured by the Canadian de Havilland plant at Downsview, Toronto, which were shipped to Chester where British equipment including radio and instruments were fitted. Although one example of the type was experimentally converted at Hatfield with the British-made Alvis Leonides engine, all the Chester-assembled Beavers retained the original standard Pratt and Whitney Wasp Junior engine.

XP918, the first of thirty D.H. 110 Sea Vixen aircraft built at Chester for the Royal Navy when the last phase of production of this aircraft type was transferred from the DH Christchurch, Hampshire, factory in 1962. After this, the last FAW Mk.1, there were twenty-nine aircraft of the FAW Mk.2 version built during the ensuing four years, incorporating overall improvements to the weapon system, together with increased fuel tank volume in forward over-wing extensions of the twin-boom structures – the last flew on 3 February 1966. A further thirty-seven Christchurch-built Mk.1s were also converted at Chester to the Mk.2 standard between 1965 and 1968. The first of these, XJ580, was the last Sea Vixen to fly in RN colours and, after a twenty-year service life, has now been donated to the Tangmere Military Aviation Museum by the Sea Vixen Society.

Originally designed and built by de Havilland at Hatfield as a jet replacement for the Dove and Heron, the D.H.125 500mph business jet was first flown there as a prototype on 13 August 1962. After the second aircraft, full-scale production was transferred to Chester to enable the Hatfield team to concentrate on the Trident airliner (first flown on 9 January 1962 but the only DH type for which Chester was not substantially involved). The first Chester-built D.H.125, G-ARYC, flew on 16 February 1963 and by mid-1964, components for the fourteenth (now re-designated) HS125 aircraft were entering final assembly, initial release to production of sixty aircraft was implemented, and transatlantic air deliveries began in July. (The folding-wing Sea Vixen naval fighter and the Comet airliner were also in continuing production at this time)

Chester-built Sea Vixen FAW Mk.2 XP924, the sixth of twenty-nine, first flown there in August 1963. Having notably served with 889 Sqn. Fleet Air Arm on the carrier HMS Eagle in the 1960s, this is the only remaining airworthy Sea Vixen. Now based at Hurn (Bournemouth), it carries the appropriate civil registration G-CVIX for display demonstrations – after restoration by de Havilland Aviation Ltd at Swansea, South Wales. The Sea Vixen was the third and largest development of the distinctive DH twin-boom jet fighter concept and the heaviest aircraft to enter British naval service. It was the first British intercepter to dispense with guns, Britain's first naval aircraft designed as an integral weapons system, and the first to become operational armed with guided weapons. In the all-weather role, it was equipped with D.H. Firestreak air-to-air infra-red homing missiles and retractable rocket pods.

The 100th HS125 business jet, a Series 1A for the US market (hence the 'D.H.' prefix because of the de Havilland name being much better-known and preferred there), leaving the Chester production line on 15 July 1966, and at which time production output was already seven aircraft a month. As the archetypal business jet, and increasingly the mainstay of the Chester output with well over 1,000 people now involved in the programme, many aircraft were being built and flown as 'green' (unfurnished) airframes – as typified by this one. Interior furnishings and custom equipment were installed by one of Hawker Siddeley's three North American distributors engaged in late 1963, Atlantic Sales Corporation of Wilmington, Delaware (the other two being AiResearch Aviation Services of Los Angeles and Timmins Aviation of Montreal, Canada). This event set a precedent for the celebration of several centennial deliveries thereafter.

An HS125 Series 3B G-AWMS built in 1969 for the RTZ (Rio Tinto Zinc) mining company at the pre-delivery flight test stage outside the former fire station and flight test offices fronting the flight shed. This model exemplified an early stage of continuous all-round development of the breed, in particular with new engines and avionic equipment, that has characterised the 125 programme from the outset of its production life to meet the needs of its diverse global appeal.

In December 1969, Hawker Siddeley joined forces with Beech Aircraft Corporation of Wichita, Kansas, as distributors of all 125s in North America and in which market the aircraft was renamed the Beechcraft Hawker BH125. Beech also agreed to purchase forty green aircraft from the Chester line and complete them as BH125s to customer requirements. This partnership was eventually severed in September 1975 when Hawker Siddeley Aviation Inc. was formed to handle HS125 sales in the USA and Mexico. (Nineteen years later, the partnership was effectively reinstated when, in 1993, Raytheon Aircraft Company, which had acquired Beech in the meantime, bought the British Aerospace Corporate Jets Co. but continued to source 125 airframes from Chester under production licence).

An HS125 Series 400A green 'airframe' destined for an unnamed American customer, but already carrying an American registration, N57BH, being readied outside the Chester flight shed in May 1971 for delivery to become a BH125. North Atlantic delivery flights were made via Iceland, Greenland and Goose Bay, Labrador, (Canada). This was the route that had been pioneered by the RAF with a six-aircraft D.H. Vampire F.3 flight in July 1948 (the first jet aircraft to cross the Atlantic under their own power). It was also used extensively by Vickers-Armstrongs from the mid-1950s for deliveries of the Viscount to North American customers.

Four of the eventual fleet of twenty HS125 Series 2s produced at Chester during the 1960s for the Royal Air Force as Dominie T.Mk.1 air navigation trainers – XW789, XW791, XX505 and XX508 awaiting delivery, initially to No.1 Air Navigation School at Stradishall, Suffolk. The RAF Dominie fleet is today located at the RAF College at Cranwell, Lincolnshire. The name was a Scottish term for schoolmaster and had originally been used by de Havilland for the pre-war D.H.89 Dragon Rapide twin-engined biplane when deployed by the RAF as a navigation trainer during the Second World War. Many other 125s were subsequently to issue from the Chester line for military applications, civil aviation pilot training and airways calibration duties.

The XV148, the second and fully representative HS801 Nimrod development prototype, in final assembly at Chester. The British Government had announced the conception of the Nimrod ('The Mighty Hunter') on 2 February 1965 as the world's first (and still the only) land-based pure jet long-range maritime reconnaissance, anti-submarine and search-and-rescue aircraft. It was based on the well-proven Comet 4 airframe, and to replace the venerable Avro Shackleton (of which a large number had been overhauled at Chester in the early 1960s).

The XR148 being made ready for its maiden flight at Hawarden on 23 May 1967. Initially completed at Chester in mid-1965 as the second of two unsold civil Comet 4Cs, and stored without being flown, this aircraft was then converted there for the Ministry of Defence, complete with the navigation and attack systems, and Rolls-Royce Spey engines in place of the original Avons.

XV814, the ex-BOAC (DH Hatfield-built) Comet 4 G-APDF which was sold to the Ministry of Technology and converted at Chester before being delivered to the Royal Aircraft Establishment (RAE), Farnborough, in October 1968. There they used it as a spacious long-range platform for military communications equipment development. Later acquiring a Nimrod fin and rudder from the first prototype, XV147, this aircraft inevitably acquired the soubriquet 'Conrod'

Several Chester-built Comet 4s were used in the complex Nimrod development programme. In 1977, the Chester-built ex-BOAC Comet 4 G-APDS became XW626, the development aircraft for the bulbous nose-mounted GEC radar for the (later abandoned) Airborne Early Warning Nimrod AEW Mk.3.

The HS801 Nimrod Mk.2 XV254, one of the RAF's current fleet of twenty-one operational aircraft (converted from Mk.1 to Mk.2 standard with up-dated avionics at HSA Woodford between 1975 and 1984) and incorporating original Chester-built main airframe components. These aircraft currently serve with 42(R), 120, 201 and 206 squadrons based at Kinloss, Morayshire, Scotland. Designed to combine high-altitude performance and fast transit speed with low wing loading and low-speed manoeuvring to provide anti-shipping strikes, submarine hunting and surveillance, the Nimrod has been in service since 1969, notably participating in the Falklands and Gulf campaigns.

Main fuselage production line for the original HS801 Nimrod Mk.1 at Chester in the late 1960s (alongside continuing high-rate production of the HS125 business jet). The two Chester-built Nimrod development prototypes, XV147 and XV148, were the precursors of all forty-one original Nimrod production aircraft fuselages, wings and other components at Chester between 1966 and 1970 (38 MR.1 maritime reconnaissance and three R.1 electronic surveillance). These major airframe components were then driven to HSA Woodford for the addition of the 'bath-tub bubble' skirt structure of the unpressurized weapons bay to the underside of the slender Comet airliner antecedent and for final assembly, equipping and delivery.

The first three of twenty-one ex-Nimrod MR2 fuselages entering the final assembly area, at what is today BAE SYSTEMS Regional Aircraft Woodford, for the second time, and now initiating the new BAE SYSTEMS Nimrod MRA4 assembly line. In July 1996, the UK Government awarded British Aerospace the contract to upgrade twenty-one Nimrod MR2s, with new wings, engines, flight deck and mission systems. Significantly, the original Chester-built fuselage structure of the late 1960s is the only major airframe component being retained, having been found to be in excellent condition on strip-down. Separated by thirty years of sterling operational service, this is an outstanding tribute to the original Chester workmanship. Moreover, this programme will extend the total service life of the DH Comet genus well into the new millennium and probably to a total time-span of between seventy and eighty years.

Following its major Nimrod prototype conversion a decade earlier, in the late-1970s Chester converted the HSA Woodford-built HS748 twin propeller-turbine airliner, G-BCDZ, repurchased from COPA of Panama, to the maritime patrol 'Coastguarder' role. It had radar housed beneath the front fuselage, bubble observation windows on each side and increased fuel capacity to provide an eleven-hour endurance. Taken to Chester on 14 May 1976, this aircraft was first flown in the new military guise on 18 February 1977 and returned to Woodford for development testing. However, appearing just two months before the formation of British Aerospace, and with no orders having been received, this singular prototype was reverted to airline configuration. Two HS748s were also transferred from Woodford to Chester for final assembly in December 1974, at the peak of Nimrod work there, and were then delivered to Germany.

The sculpture machining of rolled and forged aluminium billets to form the skin cover panels for the wing structural box/fuel tank of the Airbus A300 - the worlds' first wide-body, twin-aisle, twin-jet airliner. This was done on custom-built, numerically controlled, long-bed (60ft x 12ft air cushion/vacuum suction bed) machine tools installed in what had been the main aircraft assembly hall of the Chester factory. In 1969, Hawker Siddeley had courageously elected to become a private contract partner in the European Airbus Industrie (AI) airliner programme consortium (formally constituted in December 1970) to supply wing structures designed at Hatfield and manufactured at Chester. This had required considerable new investment in what were the largest high-speed profile-milling and routing machines in Europe, made by Marwin of Leicester and Cramic of Southall, Middlesex, to produce these 50ft long wing skin panels. A 1,200ton hydraulic horizontal-acting press was also acquired from Shaw of Salford, Lancashire, for pre-forming the access door-perforated underside panels.

Sculpture-machined Airbus A300 inner and outer wing skin panels incorporating separately-attached open l-section stiffening and stabilising stringers, fuel tank access door cut-outs and sealing compound, prior to final assembly in (four of) the six sets of upright assembly jigs in the background. By mid-1976, with orders for 100 complete wing sets, plus authorisation for a further thrty-two sets of advance materials, Chester had produced eighty wing sets, each worth more than £2million at that time.

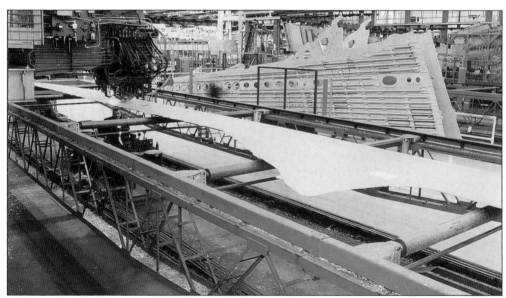

An American Gemcor Drivmatic, automated and numerically controlled, squeeze riveting machine of the type widely used in Airbus wing machined skin/stringer panel assembly at Chester since the 1970s and which have proved to be ideal for the heavy duty operation involved. These machines automatically drill and countersink the holes in the metal, form the rivets from metal blanks, fit and clench them, machine the heads flush with the wing skin outer surface and then move the whole structure to the precise position for insertion of the next rivet. The whole operation is monitored by CCTV cameras mounted above and below the structural assembly. American Spacematic and Quackenbush drill motor equipment was adopted early for the great proliferation of skin-to-structure bolt fastenings applied during the in-jig assembly operation.

The removal from the assembly jig at Chester of the first (port) Airbus A300 airliner two-part wing structure on 11 August 1971 – the first significant Chester contribution to this new pan-European commercial aircraft industrial mosaic. Moved in a specially designed lifting cradle, supported by the overhead crane rail system, this structure was an almost totally machined and integrally sealed 'wet' wing torsion box/fuel tank construction. This airliner wing manufacturing technique had already been well established in the UK by Hawker Siddeley, Vickers-Armstrongs and British Aircraft Corporation.

Out-of-jig completion of Airbus A300 wing structures with the attachment of secondary structural items prior to aerial transport first to VFW Bremen, for equipping and then to the Airbus Industrie final assembly centre at Aerospatiale, Toulouse.

Initially, the completed structures were moved up to a specially designed British Cramic machine to trim-finish the root profile to match the joint attachments on the wing/fuselage centre section, top right, but this later became unnecessary. The detail design and parts supply chain for these huge structures engaged six HSA factories, co-ordinated from the parent plant at Hatfield: Chester, Hatfield, Woodford and Chadderton (near Manchester), Brough (Yorkshire) and Hamble (near Southampton).

The first Airbus A300B wings being loaded aboard an Aeromaritime/Airbus Industrie Super Guppy freighter aircraft at Manchester Airport from a custom-designed scissor-lift trailer and bound for Bremen on 23 November 1971. This voluminous aircraft was an adaptation of the original Boeing 377 Stratocruiser transport/military tanker that was long used for the international transport of Airbus airframe components. (The name 'Guppy' derived from the fat and ungainly fish of that name). It had an outsize upper fuselage and sideways-hinging nose fuselage section and, re-engined with turboprops, followed the initial Guppy G-201 piston-engined version, originally developed by Aero-Spacelines of the USA. The original was designed for the transport of Saturn Moon rocket booster casings around the US, and in the 1970s it had been used for the aerial inter-site transport of the fully-equipped airframe assemblies for the Anglo-French Concorde supersonic airliner.

The initial Airbus A300B airliner prototype/demonstrator, incorporating Chester-built wings and registered F-WUAD, which made its maiden flight at Toulouse on 28 October 1972. Later re-registered as F-OCAZ, this major new European airliner made extensive development and sales tours before achieving certification in March 1974 to inaugurate a trail that would go on to transform the Airbus consortium. It was to change Airbus Industrie from a one-product organisation into one which today is promoting eight distinct models with numerous variants – all incorporating Chester-built wings and craftsmanship.

Seven

British Aerospace and Airbus Wings the World Over

The creation of British Aerospace (BAe) in April 1977 resulted in the Hawker Siddeley Aviation Hatfield-Chester Division becoming the similarly named division of the BAe Aircraft Group from 1 January 1978. BAe joined Airbus Industrie as a full twenty per cent equity-holding partner on 1 January 1979. The expansion of Airbus wing manufacture and continuing production of the 125 business jet were the twin staples of the Chester factory output during the ensuing twenty-two year BAe era. The closure of the long-established Hatfield parent plant in Autumn 1993 also resulted in Chester becoming conjoined with Filton (Bristol) as the divisional management and design centre of the Airbus Division of British Aerospace (Commercial Aircraft) Limited in 1989. It continued as the exclusive wing manufacturing centre for every member of the growing Airbus airliner dynasty; as well as that for the enduring 125 (with airframe component kits for Raytheon of the USA after it acquired BAe Corporate Jets in January 1993). The factory completed the 2,000th Airbus wing set in early 1999 and became known exclusively as 'Broughton', a geographically correct name in deference to Welsh devolution. Major new extensions and equipping for the newest Airbus derivatives were also begun during 1999, preceding the formation of BAE SYSTEMS on 30 November 1999.

When British Aerospace (Commercial Aircraft) Ltd. was formed on 1 January 1989, the newly-clad wing despatch building was the ideal location to proclaim the Airbus Division formed a month later. It was the loading and aerial delivery point for the finished product and the aerial arrival point of visitors and customers to the Chester site. The BAe Corporate Aircraft Division was also established at this time to handle continuing manufacturing and sales of the (now further re-designated) BAe125.

These fourteen aircraft at Chester on 23 May 1985 typify the celebrations of centennial production and/or sales events in the prolific and enduring Hawker Siddeley/British Aerospace 125 business jet programme. This event united almost every version of the aircraft with the celebration of the sale of the 600th example into a market which by then embraced thirty-seven countries world-wide with 60 per cent of the output being exported to North America. The latest version, the Series 800 in the foreground, notably offered a 3,000 nautical mile intercontinental-range and an Electronic Flight Instrument System (EFIS) equipped flight deck (the 100th sale of this version was announced on 24 October 1987). Total BAe 125 800 sales were 319 aircraft (following 576 Series 1 to 700).

The first BAe 125 Series 1000 corporate jet – registered G-EXLR, standing for EXtended Long Range (and 'Elixir' for the corresponding market invigoration) – making its ceremonial 'first flight' at Chester on 28 June 1990 (it made its actual first flight there on 16 June). It was then transferred to the BAe civil flight test centre at Woodford. This was the last BAe development of the type and all 125 production under the BAe aegis at Chester ceased in 1996 with the completion of 42 Series 1000s – and a grand total of 869 of the 125 genus. BAe then sold its Corporate Jets business to the Raytheon Company of the USA (who owned the Beech Aircraft Company to which the 125 had been linked nineteen years earlier) on 1 June 1993. Renamed Raytheon Corporate Jets Inc (RCJ), with the permission of Hawker Siddeley, the aircraft were henceforth marketed as 'Hawkers' An up-graded version of the BAe 800 was renamed the Hawker 800XP (Extra Performance) with airframe component kits continuing to made at Chester before being shipped to Wichita for final assembly.

The Raytheon Aircraft Services Limited (RASL) stand-alone, management and twin-hangar (originally RAF MU K-type) facility to the north-east of the Broughton site. While the ex-BAe corporate jet assembly line was moved to Raytheon's Little Rock, Wichita, plant, BAe was contracted to continue to build the airframe structure kits (but moved to a separate and discrete area of the factory site). Two years later, Raytheon made what had been the BAe125 service centre a stand-alone service centre for its 'Hawkers' operating in Europe, Africa and the Middle East, and in 1997 fully integrated it into the Raytheon Aircraft Services global network.

A Bombardier Global Express aircraft visiting the Raytheon Aircraft Services Centre at Broughton for modification, systems integration and test flying as part of the development of the air segment of the ASTOR (Airborne Stand-Off Radar) system for the UK Ministry of Defence – of which five examples are being procured. In 1998, Raytheon began extending its capabilities at the centre to include this kind of specialised conversion work and to build on its high skill base and that of the local area.

Airbus Wing Manufacturing Milestones

The progressive build-up of Airbus wing manufacture – through the thirty-year Hawker Siddeley/British Aerospace/BAE SYSTEMS continuum – to meet both the increasing number of customer orders and the continually diversifying Airbus Industrie (AI) airliner family has paralleled the evolution and global success of the whole AI enterprise. All twin-aisle aircraft wings (A300/A310/A330/A340) have been delivered to the German Airbus manufacturing centre Bremen for equipping, before onward air transport to the AI final assembly at Toulouse. The first wing set for the single-aisle regional A320 was despatched from Chester to BAe Filton, Bristol, on 28 November 1985 to a new Wing Equipping Centre there, before onward transport to Toulouse. This continued until July 1993 with the completion of the 453rd A320/A321 wing set, whence the equipping of these wings was reverted to Chester, before despatch to the AI final assembly centres at Toulouse and Hamburg respectively (and since early 1995 also including the A319 for the Hamburg line).

1971 August	*First A300 wing box out-of-jig.*
1971 November	*First A300 wing set despatched to VFW in Bremen for equipping.*
1979 June	*Delivery of the 100th A300 wing set to Bremen.*
1981 January	*150th Airbus wing set delivered.*
1981 May	*First A310 wing box delivered.*
1985 July	*Assembly of the first A320 wing started.*
1985 September	*First A320 wing removed from the assembly jig.*
1985 November	*First A320 wing set leaves for BAe Filton for equipping, testing and completion*
1989 August	*New £5 million 'link' building (between the original main assembly area and the former flight shed) commissioned to accomodate increasing production.*
1989 August	*500th set of A300/A310 wings despatched to Bremen (329 A300s and 171 A310s).*
1990 June	*First A340 wings removed from assembly jig and despatched to Bremen.*
1991 February	*200th A310 wing set despatched.*

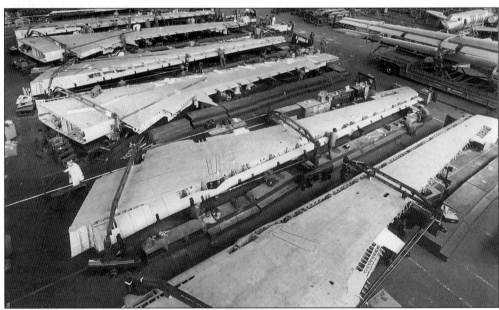

Out-of-jig completion of Airbus A300 and A310 wing structures in the early 1980s.

1991 August	First A330 wing set removed from the assembly jig.
1991 September	New £3.2 million Wing Despatch Centre opened and first A330 wing despatched.
1992 March	First A321 wing removed from assembly jig.
1993 March	First phase of transferring A320/A321 wing equipping from Filton to Chester started.
1993 July	453rd and last wing set equipping completed at Filton.
1994 August	Work begins on the 100th A330/A340 wing set.
1994 September	First metal cut for the A319 wing.
1994 November	100th A330/A340 wing set delivered.
1994 December	500th A320 wing set delivered.
1995 October	First loading of an A340 wing set into the A300-600T SuperTransporter 'Beluga' aircraft.
1999 February	2,000th Airbus wing set delivered.
1999 April	British Aerospace Airbus receives the Queen's Award for Export Achievement 'in recognition of the significant contribution that the export of Airbus wings makes to the UK economy'.
2000 September	Delivery of the 2,500th Airbus wing set (i.e. 5,000 individual wing structures).
2000 September	Decision announced that A3XX wings to be built at Broughton.
2000 December	A3XX renamed A380 with full programme launch.

A major Airbus wing production milestone being appropriately celebrated in May 1992 with the hand-over of the 999th wing set, for an A340 for Lufthansa of Germany, in front of an undesignated A321. The 1,000th wing set, for an A310 for the Oasis Group of Spain and the 1,001st, for an A320 for Northwest Airlines of the USA, were also handed over in similar fashion at the same time. The whole triple celebration was held in the presence of a thousand members of the Chester workforce. they were arranged to emphasise visually the British Aerospace 'Arrow' corporate logo – thereby symbolising both the explosive growth of the Airbus family and its geographical market penetration, and justifying the contemporary and proudly proclaimed BAe slogan: 'Airbus – A Great British Success the World Over'.

Celebrating the emergence of the first wing box structure for the Airbus single-aisle A320 from the assembly jig on 3 October 1985 – three weeks ahead of schedule. Both wing boxes then left Chester by road on 28 November – prominently labelled as such and placarded 'Beware of Fast Mover' – for transport to the BAe Airbus Division Filton (Bristol) site, arriving there the same day, for equipping prior to despatch to Toulouse for final assembly.

Equipping the Airbus A320 family (A320/A321 A319/A318) wings with secondary and wing tip fixed structures, leading and trailing edge flying control surfaces and fuel and de-icing systems at the Single-Aisle Wing (SAW) equipping centre at Chester.

A completed 98ft/30m wing torsion box structure for the Airbus A340 after removal from the multi-level assembly jig – by an inflatable air-bag 'hovercraft-style' raft. After initially being transported in pairs by Super Guppy aircraft to Bremen for equipping, they are flown on to Toulouse singly for final assembly because of the consequent extended dimensions in relation to those of the upper fuselage cross-section of both the Super Guppy and the succeeding Super Transporter aircraft. The A340 was the first four-jet long-range derivative of the Airbus family, with a range of up to 15,000km, and has been flying in scheduled service with Chester-made wings since 1993.

One of the two boldly-titled wing boxes for the first Airbus A340 leaving BAe Chester by road in June 1990 for aerial transport by Super Guppy aircraft from Manchester Airport to Bremen for equipping and emphasising the huge scale of both the structure and the custom-designed 115ft-long transporter vehicle.

One of the fleet of five specially-designed Airbus A300-600ST Super Transporter aircraft loading an Airbus A320 wing set from a custom-built laser-aligned elevating platform via the clam-shell nose-loading door of the main deck above the low-mounted flight-deck. Based on the A300-600R, this is the world's most voluminous cargo aircraft and was specifically developed and manufactured for Airbus by the Special Aircraft Transport International Company (SATIC), a joint subsidiary of the French Aerospatiale-Matra Airbus and German DaimlerChrysler Aerospace Airbus. It has been progressively replacing the ageing Super Guppy aircraft from January 1996. Nicknamed 'Beluga', due to its resemblance to the great white whale of that name, the 7.4m wide cabin has a 1,400cum volume and 45.5 tonne capacity. This volume capacity is sufficient to transport one set of wing boxes for the A300/A330/A340 or two sets for the A320/A321/A319/A318 or single wings for the latest extended wing A340-500/600 series. These aircraft now visit Broughton every working day, with turnaround times of approximately one hour, for the rapid transport of completed Airbus wing sets to the respective continental final assembly centres in France and Germany as appropriate.

The 1,500th set of Airbus wings being loaded into an Airbus Super Transporter aircraft at the BAe Chester factory in August 1996, destined for the German Airbus assembly centre at Hamburg. This wing set was for an A319 – the smallest Airbus type at that time – and appropriately part of an order for the German airline Lufthansa. At that time, BAe stated that 300 other companies and 25,000 workers in Britain were involved in Airbus design and manufacture and that throughout the UK Airbus business accounted for nearly 1.5 per cent of total national manufacturing exports and was contributing £1 billion to the annual trade balance.

The new 100ft long wing skin panel milling machine made by MPS of Leicester for the extended span Airbus A340-500/600 series, installed during 1999, the largest and longest to date. Those installed at Chester in the early 1970s to produce the 50ft-long panels for the two-part Airbus A300 wing were the largest in Europe, until in 1984 the world's largest routing machine was brought into operation, the machine bed measuring 66m x 3.75m and weighing 532 tonnes. The cutting heads of the KTM 200 machines that were installed in 1990 were also fully overhauled and up-dated by Giddings and Lewis of Southampton in 1999.

The new vertical wing spar milling machine built by Ingersoll of Rockford, Illinois, USA, with vertical part holding, linear motors and large flood coolant tanks, installed in 1999. The Broughton wing skin and spar machine shop is complemented by a major new anodising process treatment and painting facility also commissioned during 1999. Fully-automated, the twelve-tank facility includes a washing, degreasing and rinsing line to cleanse incoming components, followed by deoxidisation and chromic acid processes, before air-drying and painting. With computer-controlled cranes, each with a maximum capacity of four tonnes to manoeuvre components through the process line, this huge facility can process thirty-four flight bar loads per day.

Two of four large-scale 'state of the art' electro-impact Low Voltage Electromagnetic Riveter (LVER) machines installed in 1999 to enhance the capability of automated panel assembly with precision drilling and insertion of the many large-size bolt and rivet fasteners which attach the machined strengthening stringers to the wing panels. Modelled on the Airbus A320 wing skin LVER riveter, these new machines incorporate faster insertion speeds and a 'cold-working' facility into the process. Each set of single-aisle aircraft wings, the smallest produced, requires around 10,000 rivets and 16,000 bolts.

Eight
BAe Systems And Airbus UK Into The New Millenium

Broughton (now the exclusive sight name) entered the new millennium under the banner of the newly formed BAE SYSTEMS – which was created on 30 November 1999 by the merger of the former British Aerospace (BAe) and GEC-Marconi Electronic Systems (MES). Broughton then became part of Airbus UK Ltd, formed by BAE SYSTEMS on 30 March 2000 in anticipation of the announcement of the long gestated pan-european Airbus Integrated Company (AIC), which came on 23 June 2000. The new pan european became operational from 1 January 2001 and within which Broughton continues to operate integrally with the Airbus UK management, design and manufacturing centre at Filton (Bristol). Spurred by the courage and value of the vindicated Hawker Siddeley decision to become a private venture partner to the fledgling Airbus consortium thirty years ago, and massively developed through the twenty-two-year British Aerospace regime, Broughton is universally acclaimed as one of the world's most efficient and productive aerostructure producers. With well over £100 million of fresh investment in design and manufacturing facilities in 1999, Broughton is now preparing to meet the challenge of the wings for the new Airbus A380 double-deck airliner. This healthy situation has provided a fitting Diamond Jubilee tribute to the great heritage, capability and reputation that Broughton bequeaths to the globally successful Airbus family, founded by its forebears six decades ago with the nation-saving Vickers Wellington.

The in-jig upper-platform assembly of the main wing torsion box structure of the Airbus A340-50D/600 series – the largest version in the current Airbus production range. It has integrally end mill-routed chord-wise ribs, span-wise spars, fixed leading-edge slat support brackets, machined bottom skin cover panels and separately-machined and attached stringers – prior to attachment of the top skin panel/stringers. The whole wing box assembly, incorporating the outer section made at Filton, then forms a sealed integral 'wet wing' fuel tank. This general constructional concept has characterised the entire Airbus wing family to date.

The BAE SYSTEMS *Airbus* UK *Broughton factory and Hawarden airfield, August 2000. A £100 million capital investment programme in 1999 resulted in large-scale 'millennium generation' machining, forming and treatment facilities; the corresponding buildings and infrastructure are complemented by expanded wing box assembly and equipping areas. The new facilities include: a A340-500/600 wing skin milling facility and production line; four Low Voltage Electromagnetic Riveter (LVER) machines; a stringer manufacturing centre; a large-scale wing spar mill; in-house wing skin, spar and stringer forming plant; automated twelve-tank treatments facility; and long-range wing equipping facility. The main new building, downwards from upper centre right: the twin-aisle aircraft wing equipping centre (beginning with the extension at the top right corner of the administration block and main factory); the separate (transverse) single-aisle aircraft wing equipping building; and the Metal Improvement Company (MIC)* building replacing the former Deeside plant. (The £5 million 'link' building between the main factory and the former flight shed was commissioned in August 1989). All of this development is providing a world-class 'centre of excellence' for long-bed machining, processes and handling for the latest 'flow-line' manufacturing techniques for wing structures for the full range of current and future Airbus airliners (cf. the similar view of the site in July 1947 on page 34.)

Roll out at Toulouse of first A340-600 on 23 March 2001.

The huge scale is evident of the first 100ft long A340-600 port wing emerging from the assembly jig in mid-April 2000. The largest and highest-technology Airbus wing built at Broughton to date, it was transferred by overhead crane and custom-built lifting cradle to the new Twin Aisle Wing Equipping Centre, for the installation of secondary structure, fuel and other systems. It was then despatched singly to Bremen (not as a pair as with the all other Airbus type wings) in a suitably-modified Super Transporter aircraft for final equipping with leading- and trailing-edge control surfaces.

Impression of the diminutive 107-seat Airbus A318, the smallest member of the Airbus airliner range to be committed to production. This third derivative of the highly successful 150-seat A320 single-aisle 'reference standard' regional jetliner family will complement the 185-seat A321 and 125-seat (and Corporate Jet) A319 that are well-established in production and service. The first set of wings for the A318 is due for completion at Broughton in August 2001 and the new type is scheduled to begin airline service in the third quarter of 2002. Launch customers notably include British Airways (BA).

Enduring Legacy And Millennium Tribute

The Avro Lancaster PA474 of the RAF Battle of Britain Memorial Flight (BBMF) is a reminder of Vickers-Armstrongs' prodigious bomber production at the wartime Shadow Factory at Broughton. First built there in mid-1945, the Lancaster is part of Broughton's rich six-decade legacy to BAE SYSTEMS and Airbus UK. This aircraft is displayed annually throughout the United Kingdom, a tribute to both the Broughton wartime output and its own classic status. The best known of 7,377 Lancasters built in the UK and Canada, 235 of which came from Broughton, PA474 is one of two surviving airworthy examples in the world (the other is in Canada). Intended for service in the Far East, shortly after it was delivered the abrupt end of the war with Japan led to its assignment to photo reconnaissance duties with No.82 Squadron RAF in East and South Africa. It was then transferred to the College of Aeronautics at Cranfield, Bedfordshire for flight trials with a Handley Page laminar-flow wing (mounted vertically above the rear fuselage) until 1964. Afterwards adopted by the Air Historical Branch (AHB), it appeared in two well-known films – *Operation Crossbow* and *The Guns of Navarone*. The markings of No.44 Squadron RAF reflected the fact that this squadron became the first operational unit to receive the Lancaster in 1941 at Waddington, Lincolnshire. PA474 was transferred there in 1945, and given the markings L5758 'KM-B', commemorating John Nettleton VC and the aircraft that he flew on the raid on Augsberg on 17 April 1942.

After restoration, the aircraft joined the BBMF in November 1973 and two years later was adopted by the City of Lincoln. During the winter of 1995/96 it received a new main wing spar and the markings of W4964 'WS-J', *Johnny Walker*, an aircraft of No.9 Squadron RAF which took part in the first attack on the German battleship *Tirpitz*. Undergoing a major overhaul at RAF St. Athan, South Wales in the winter of 1999/2000, the aircraft was again repainted, this time in the markings of No.61 Squadron EE176 'QR-M' with the nose art 'Mickey the Moocher', Walt Disney's Mickey Mouse cartoon character (but still carrying its original RAF serial PA474). It now represents one of Bomber Command's 'centenarians' – aircraft which took part in 100 or more wartime operations – and it continues to lead the renowned BBMF fly-past from its home base at RAF Coningsby, Lincolnshire.

Crown Copyright/MOD. Reproduced with the permission of the Controller of Her Majesty's Stationery Office.

One of the more than a thousand de Havilland/Hawker Siddeley/British Aerospace 125, and now Raytheon Hawker, corporate business jets built at Broughton since 1962 – being fettled in the Raytheon Aircraft services centre there. The World's first jet aircraft specifically designed as a corporate communications aircraft, this diminutive world-beater has been an enduring highlight and mainstay of the prolific production output of the Broughton factory for nearly thirty years. A prime export earner for Britain, the North American dollar market has accounted for around two-thirds of total sales.

Impression of the new '21st Century Flagship' Airbus A380 480-650 seat double-deck 'Super Jumbo' airliner and 150-tonne freighter (the world's largest). It is intended to fill the only area of the Airbus range in which the rival Boeing 747 does not have a direct competitor. In April 2000 the UK Government made available a £530million repayable launch investment and in September 2000, the Welsh National Assembly granted a £19.5million support package which also confirmed that the wings are to be built at Broughton. Full programme launch of the A380 (hitherto known as the A3XX) was announced on 19 december 2000. This huge aircraft is designed within an 80m square to match existing airport aprons and taxiways and will incorporate a considerably enhanced wing and high-lift system to enable it to use existing runways. First deliveries of the A380 are scheduled during 2006.

The Broughton Production Record

Type	Time Period	No. Built
Vickers-Armstrongs		
Wellington Mk.I	1939	3
Wellington Mk. IA	1940	17
Wellington Mk. IC	1940-1942	1583
Wellington Mk. III	1941-1942	737
Wellington Mk. IV	1940-42	220
Wellington Mk.X	1942-1944	2434
Wellington Mk.XII Tropicalised	1943-1944	8
Wellington Mk.XIV	1943-1944	538
Avro Lancaster B.Mk.1	1944-1945	235
Avro Lincoln	1945	11
Total:	1939-1945	5786
de Havilland		
Mosquito	1948-1950	81
Hornet	1948-1952	149
DHC Chipmunk	1950-1956	889
Vampire J.28B (Sweden)	1949-1952	297
Vampire F.B.Mk.5	1951-1952	87
Vampire F.B.Mk.9	1950-1956	255
Vampire N.F.Mk.10	1951-1952	55
Vampire F.B.Mk.52	1950-1953	114
Vampire N.F.Mk.54	1952-1953	12
Vampire T.11 Trainer	1952-1956	260
Vampire T. 55 Export Trainer	1955-1963	164
Venom F.B.Mk.1	1952-1955	302
Venom N.F.Mk.2	1952-1955	139
Venom N.F.Mk.3	1955-1956	83
Venom N.F.Mk.51/J.33 (Sweden)	1952-1957	62
Venom F.B.Mk.4	1955-1956	73
Venom F.A.W.Mk.20	1954-1955	38
Venom F.A.W.Mk.21	1955-1956	99
Venom F.A.W.Mk.22	1956-1958	38
Dove	1951-1967	244
Heron	1953-1967	140
Comet 2	1957	1*
Comet 4	1958-1964	40
DHC Beaver	1960-1967	46
125	1962-1963	4
Sea Vixen	1962-1966	30
* Four others only partially completed		
Total:	1948-1966	2816
Hawker Siddeley		
125 Series 1-700	1962-1977	380
Nimrod Development prototypes	1965	2
Total:	1962-1977	382
British Aerospace		
125 Series 1-700	1978-1985	193

125 800	1983-1996	252
125 1000	1990-1996	42
Total:	1978-1996	487
TOTAL 'FLY-AWAY' AIRCRAFT BUILT	1948-1996	9471

Airbus Wing Sets:
Hawker Siddeley

A300	1971-1977	45
Total:	1971-1977	45

British Aerospace

A300	1977-1999	442
A310	1981-1999	255
A320	1985-1999	812
A340	1990-1999	148
A330	1992-1999	224
A321	1992-1999	152
A319	1995-1999	180
Total:	1977-1999	2213

BAE SYSTEMS and Airbus UK

All Airbus types	1999-March 2001	453
TOTAL AIRBUS WING SETS November 1971 - March 2001		2666

Other Work Programmes:
Vickers-Armstrongs

Pre-fabricated Houses	1945-1948	11,250

de Havilland

Dove and Heron overhauls		
Royal Flight Heron servicing		
RCAF Comet I modifications	1956	
Avro Shackleton overhauls	1960-1962	
RAF Comet C2 strengthening		
Vampire refurbishment	1954-1964	

Hawker Siddeley

Trident component manufacture	1960s	
Blackburn Beverley overhauls	1962	
Ex-BOAC Comet 4 modn for RAE	1963	1
Vampire Trainer overhauls/resale	1964-1968	87
Sea Vixen conversions	1965-1968	37
Nimrod main airframe components	1966-1970	41
Hawker Hunter overhauls	1971	
HS748 final assembly	1974-1975	2
HS748 Coastguarder conversion	1976	1

British Aerospace

Raytheon Hawker 800XP kits	1996-1999	147

BAE SYSTEMS and Airbus UK

Raythoen Hawker 800XP kits	1999-March 2001	139

Acknowledgements

I have been privileged to know and work with most of the Vickers-Weybridge people who originally set up the Broughton Shadow Factory, and many others during the de Havilland, Hawker Siddeley and British Aerospace tenures of the plant. I must express my profound gratitude to them, and the thousands like them, for the magnificent story they created, which continues happily in high activity, and that I have retold pictorially here.

To do so has required the help of many good friends and other willing supporters in loaning and providing appropriate photographs and information which I much appreciate. I am equally grateful to them for preserving this precious material in the first place, despite the many vicissitudes through which much of it has inevitably passed over the intervening sixty years.

In this respect, I am profoundly thankful to several good friends at BAE SYSTEMS: especially for the exceptional efforts of Mark Challoner at Airbus UK Broughton. Also to Ron Hedges, Barry Guess and Mike Fielding at BAe Heritage, Farnborough; to Howard Berry, Dave Charlton and Mike Fish at Airbus UK Filton; and to Catherine Holmes and Jamie Spoors at Broughton. I must also thank Francoise Tripiano at Airbus, Toulouse, and Mike Brown of Raytheon Systems Ltd in London.

I am likewise most grateful to my old Vickers friends, Hugh Scrope, for his knowledge and records of the original Vickers-Armstrongs Shadow Factory, and Eric Morgan for the loan of the valued photos of its construction and early operation – also to John Wells and Graham Waller for access to the Vickers PLC archives now carefully preserved at the Cambridge University Library.

My special thanks also go to Phil Jarrett, whose photographic collection covering the whole of the Broughton period has proved invaluable. Also to Barry Abraham for his expert historical knowledge of the various airfields referred to and his many helpful comments. Others who have been most helpful include Philip Birtles, Norman Boorer, Geoff Green, Harry Holmes, Ian Lowe, Alec Lumsden, Sir Peter Masefield, Arthur Rowland, David Smith (especially in respect of his privately-published book 'Hawarden - A Welsh Airfield 1939-79'), Mike Stroud, Len Warrey, Les and Michael Webb, and colleagues at Brooklands Museum.

I must also thank Christine Gregory at the RAF Museum, Hendon; Sandy Gilbert-Wykes at the Battle of Britain Memorial Flight, Derek Elliott at the Central Register for Air Photography at the National Assembly for Wales; and Elizabeth Pettitt at Flintshire County Council Records Office.

Lastly, I must express my fullest appreciation of the original inspiration and encouragement of my dear old friend, the late Jeffrey 'Monty' Montgomery, in tackling this intensely interesting and deserving subject and, together with his sister, the late Diana Amberton, for first-hand memories of their time at Broughton with their parents during the profoundly challenging formative years of the Broughton saga.

Dr. Norman Barfield
Weybridge

2001